SCIENCE FOR ALL SEASONS

SHEILA KYSER BLACKWELDER is currently an adjunct instructor at Dekalb Community College. Recently she has been presenting science workshops to teachers in public school districts as well as to teachers in private schools and organizations.

SCIENCE FOR ALL SEASONS

SCIENCE EXPERIENCES FOR YOUNG CHILDREN

SHEILA KYSER BLACKWELDER

PRENTICE-HALL, INC., Englewood Cliffs, New Jersey 07632

Library of Congress Cataloging in Publication Data

Blackwelder, Sheila Kyser.
Science for all seasons.

(A Spectrum book)
Bibliography: p. 255–59
1. Science—Study and teaching (Primary) I. Title.
LB1532.B54 372.3′5′044 80-12518
ISBN 0-13-795286-4
ISBN 0-13-795278-3 (pbk.)

Editorial/production supervision and interior design by Suse L. Cioffi
Cover design by Michael Aron
Manufacturing buyer: Cathie Lenard

A SPECTRUM BOOK

10 9 8 7 6 5 4 3 2 1

Printed in the United States of America

Prentice-Hall International, Inc., *London*
Prentice-Hall of Australia Pty. Limited, *Sydney*
Prentice-Hall of Canada, Ltd., *Toronto*
Prentice-Hall of India Private Limited, *New Delhi*
Prentice-Hall of Japan, Inc., *Tokyo*
Prentice-Hall of Southeast Asia Pte. Ltd., *Singapore*
Whitehall Books Limited, *Wellington, New Zealand*

CONTENTS

PREFACE

Why is the sky blue? What makes the stars twinkle? What is a rainbow made of? Will the thunder hurt me? These or similar types of questions are asked daily of adults who work with young children. The children need to satisfy a natural curiosity about their world, so they continually ask questions. Many educators feel that young children are not capable of comprehending the complicated subject matter, such as physics or meteorology, that is necessary for understanding the scientific answers to their questions. Based on the research of Jean Piaget, educators have generally considered the young child's thinking too immature for the study of science.

Research confirms Piaget's theory that a child's thinking follows an ordered predetermined path that cannot be changed. However, there is research that indicates what teachers and parents have long suspected—that young children reach a stage of logical thinking earlier than previously believed and that the process of logical thinking may be accelerated.[1]

[1] Willard W. Hartup and Nancy L. Smothergill, eds., "Finding the Clue to Young Children's Thought Process," in *The Young Child: Reviews of Research* (Washington: National Association for the Education of Young Children, 1967), pp. 89-90.

Although there is little reason for pushing children by accelerating the acquisition of logical thinking, there is also no valid reason for restraint in the area of cognitive development. Jerome Bruner hypothesizes that any subject matter can be taught effectively in some intellectually honest form to any child at any stage of development.[2] If this is true, it seems that the time to introduce scientific information occurs when the child has a strong interest and when questions about scientific phenomena arise. There is little disagreement among educators that interest in the subject matter is the most effective teaching device known.

Educators are also in full agreement that young children learn through activity.[3] This is well expressed by a popular quotation used by many educators: "I hear and I forget . . . I see and I remember . . . I do and I understand." It seems, then, that the sensible way to begin planning a science program for young children is with activity—or "hands on" learning situations. Activity stimulates interest as well as the senses and helps to improve motor control, eye-hand coordination, language, and thinking skills.

In order to aid young children in reaching a stage of cognitive development in which they are capable of logical thinking, we should provide activities that foster intellectual growth. Children need experiences in observing, identifying, and classifying. Classification should include varying stages of complexity, from simple classification (grouping objects based on one characteristic) to multiple classification (grouping objects based on more than one characteristic in order to determine that one object is capable of being classified in several ways).

Other prerequisites for logical thinking include seriation (ordering, ranking, or creating a sequential arrangement) and conservation (understanding the fact that a whole remains unchanged

[2] Jerome S. Bruner, *The Process of Education* (Cambridge, Mass.: Harvard University Press, 1960), p. 33.

[3] Millie Almy, *Young Children's Thinking: Studies of Some Aspects of Piaget's Theory* (New York: Teachers College Press, Columbia University, 1966), p. 137.

in mass, volume, or quantity, or weight despite how its elements may be rearranged). Conservation, it seems, cannot be taught, but tasks leading to the acquisition of conservation include measuring length, height, and distance; counting and determining correspondence (as many as, more than, less than); and counting by regrouping (counting the number of coins arranged in a line and rearranging the coins in order to determine whether or not the number is the same). Activities involving class inclusion (all dogs are animals, but not all animals are dogs) and temporal relations (time and how the child perceives it) also help lead to logical thinking.

Other skills fostering intellectual development include verbalization and communication. A young child's immature thinking can lead to misconceptions that block logical thinking and can create erroneous information out of what may appear to an adult to be the most logical visual evidence. The solution to this situation may be, in part, the encouragement of verbal communication and open discussion among the children to clarify what they see in their activities performed under careful and selected adult guidance. Since too much interference from the adult stifles discovery, the adult's task is to stimulate interest by providing manipulative materials and to help define an activity more clearly in order to eliminate confusion.

The material in this book is designed to encourage discovery and the use of problem-solving skills. Although the units are designed to help children to discover answers to specific questions, the final answers are no more important than the processes by which they are derived. Nor are they more valuable in the long run than the encouragement of a feeling of success in the child. Discovering how to learn is as important as what is learned.

ACKNOWLEDGMENTS

The author wishes to acknowledge the help of the following people who assisted in the preparation of this book by reviewing the content for scientific accuracy:

Stan Adams and associates, United States Forest Service, Atlanta, Georgia.

Jim Buckley, Meteorologist, Fernbank Science Center, Atlanta, Georgia.

Nelson Cooley, Ph.D., Microbiologist, United States Environmental Protection Agency, Environmental Research Laboratory, Gulf Breeze, Florida.

Kelley Dopson, Medical Student, Medical College of Georgia, Augusta, Georgia.

David Dundee, Astronomer, Fernbank Science Center, Atlanta, Georgia.

Bernice Jordan, Biology Department Chairman, Fernbank Science Center, Atlanta, Georgia.

Fred Sherberger, Ph.D., Entomologist, Fernbank Science Center, Atlanta, Georgia.

Harold Tiller, Physicist, Fernbank Science Center, Atlanta, Georgia.

For their assistance in testing, reviewing, and criticizing this material, I am grateful to James M. Blackwelder, Mary Hollis, and Cecilia Waddey. My gratitude is also extended to Ann J. Critchfield for her work in preparing the illustrations.

INTRODUCTION

All topics in this book are presented through a seasonal approach, which is a logical way of presenting science material to young children because it allows the adult to help the child discover answers to questions about science when the time is right, when the questions are most likely to arise, and when the child experiences the phenomenon in his or her environment. A seasonal approach also illustrates the variety of science subjects that may be presented throughout the year. Science, however, does not have to be a subject presented as an entity in itself; it may be integrated into the classroom curriculum or home environment to enrich studies already taking place.

The seasonal approach is meant to be merely a suggestion for scheduling science activities. In many cases, the material may be presented whenever it seems appropriate and may enhance other learning situations. For example, the units dealing with the body (Chapter 1) are placed in this particular chapter in order to coordinate them with the study of self-awareness, which is often presented at the beginning of the school year. However, these units may be used successfully as a group to enhance other areas of study, such as health or community helpers. An individual unit on the body may enhance a variety of learning situations: For example, "Bones: The Skeleton" may be presented in October and a human skeleton compared to a Halloween skeleton; "The Heart and Circulatory System" may be presented in February and the human heart compared to a valentine heart; and "The Digestive System" may be coordinated with a study of foods and nutrition.

This material is designed for children from ages 4 through 8, although many of the units have been tested successfully with younger and older children by teachers as well as by parents not trained in the teaching profession. Those units dealing with visual analogies, those involving the body, for instance, may be used more successfully with older children. The units dealing with observation, identification, classification, and seriation may be used successfully with any age group for which the book is designed.

It is preferable that the children (especially the younger ones) be allowed to participate or not participate in the activities as they desire. This will allow each child to extract the knowledge for which he or she is developmentally ready. The range of involvement in a single group of children may vary from grasping the complete concept and reaching the answers to the proposed questions to merely enjoying and benefiting from the manipulation of the materials. This variation is unimportant, as the children will use the knowledge they have obtained to build toward complete concept development at a later time.

For the activities to be successful, the children need a working knowledge of certain basic concepts and adequate language skills before beginning the units. Teachers or parents, particularly those working with preschool children, will need to determine the level of the child's background. This may be achieved by having the children perform simple tasks in which they are asked to demonstrate a knowledge of the necessary concepts and language skills. The following list offers examples of the areas in which the children need to demonstrate proficiency:

Language concepts

hot-cold	wet-dry	big-little	dark-light
liquid-solid	heavy-light	high-low	deep-shallow
hard-soft	full-empty	opened-closed	many-few
alike-different	loud-quiet	rough-smooth	long-short

Positions

under	above	near	below
in front of	in	far from	middle
behind	out	through	between
on top of	next to	over	in back of

Skills in identification

color	shape	texture	size
quantity	sound	odor	taste

HOW TO USE THE BOOK

Information for Adults

Each unit begins with an introduction directed to the adult using the material. This information is intended to give the adult a basic working knowledge for presenting the unit and for answering the child's questions about the subject matter. For the adult who wishes to cover the subject matter in more detail, a bibliography has been included at the end of the book.

Materials

Included for each experience is a list of necessary materials. These are inexpensive items found in the home or classroom or easily obtained or constructed. The materials are the most important part of the unit, since they provide the "hands on" activity for the children. Unless it is specifically stated that the adult should manipulate the materials (usually for safety reasons), the children should be allowed to perform the activities and manipulate the materials with as little adult interference as possible.

Experience

Brief directions for performing the activities are given under the heading "Experience." It is not necessary for an entire unit to be presented in one day or one week. Let the children determine the speed at which the material is covered and let their interest determine the depth of coverage. Most of the individual experiences may be presented alone without presenting any other experiences. Any exceptions to this will be specified in the text.

Suggested Questions

In recognition of the importance of placing details into a structured pattern, suggested guiding questions are provided at the end of each experience. These are intended to guide the children through the activities, clarifying information as they move toward the final development of the concept. Some of the questions may be used as an introduction to stimulate interest in a unit which has not been child prompted. For adults interested in a means of evaluation, the questions may also be used for this purpose. The questions include those dealing with basic scientific facts, evaluative questions that require the children to make value judgments, synthesis questions that require the children to combine factual elements from several situations in order to arrive at a final solution, and divergent questions that stimulate creative thinking.

Follow-Up Activities

The final section in each unit includes follow-up activities. These provide the children with opportunities to reinforce the learning situation so that the experience does not become an isolated and perhaps forgotten event. Included as a follow-up is a list of suggested children's books.

Techniques used in these units have been borrowed from many philosophies, and it is expected that each adult using the material with children will add other variations. The combination of the manipulative activities, organized to encourage discovery, and the guiding questions that add some structure to the activity places these experiences into a category that might be best described as "guided discovery."

References

Almy, Millie. *Young Children's Thinking: Studies of Some Aspects of Piaget's Theory.* New York: Teachers College Press, Columbia University, 1966.

Bruner, Jerome S. *The Process of Education.* Cambridge, Mass: Harvard University Press, 1960.

Holt, John. *How Children Learn.* New York: Pitman Publishing Corporation, 1967.

Hunt, J. McVickers. *Intelligence and Experience.* New York: Ronald Press, 1961.

Montessori, Maria. *Childhood Education.* Chicago: Henry Regnery, 1974.

Roeper, Annemarie, and Sigel, Irving. "Finding the Clue to Children's Thought Process." In *The Young Child: Reviews of Research,* edited by Willard W. Hartup and Nancy L. Smothergill. Washington: National Association for the Education of Young Children, 1967.

Schweble, Milton, and Raph, Jane. eds. *Piaget in the Classroom.* New York: Basic Books, 1973.

1

FALL

AIR AND THE LUNGS

Information for Adults

Every living thing needs air. Nitrogen, oxygen, and carbon dioxide make up the air, which exists almost everywhere. Animals take oxygen from the air and give off carbon dioxide, whereas plants take carbon dioxide from the air and give off oxygen.

The system in the body responsible for supplying oxygen is the respiratory system, which consists of the lungs, nose, mouth, and a pathway that may be called the windpipe leading from the nose or mouth to the lungs. This pathway begins as the trachea, where air is taken in through the nose or mouth, and divides into two tubes, called bronchi, as it enters the chest and nears the lungs. Inside the lungs, the bronchi branch out to distribute the air.

In the lungs, air enters tiny sacs called alveoli. Here, oxygen is taken in, and carbon dioxide is given off. These air sacs cause the inside of the lungs to resemble a sponge. Blood flowing around the alveoli takes on a supply of oxygen to be distributed to all parts of the body and gives up carbon dioxide almost simultaneously. For the children, this may be compared to a delivery truck delivering soda pop to the store, picking up used bottles, and returning to the factory to drop off the used bottles and to pick up a fresh supply of soda pop to be distributed again to the stores.

Experience: Part I

PURPOSE: To discover that air exists all around us even though we cannot see it.

MATERIALS: Clear plastic bottles that may be squeezed; one balloon; small plastic sandwich bags (one bag for each child).

Show the children a clear plastic bottle and ask whether they see anything in the bottle. Hold the bottle near a child's face and squeeze it in order to force the air out. Discuss whether the bottle was actually empty or whether the child felt something escaping from it. This may be repeated with a variety of bottles.

Blow up the balloon as the children watch. Let them feel the balloon in order to establish that there is something in it. Allow the air to escape while pointing the balloon toward a child's face. Discuss what this "thing" is that we can feel but cannot see. If necessary, tell the children that this is air.

In order to establish that air exists almost everywhere, give each child a plastic sandwich bag and have them move around the room "catching air." Have them feel the air inside the bag, and then squeeze the bag to force the air out.

Suggested Questions

1. What is there inside the plastic bottles and inside the balloon that can be felt but cannot be seen?
2. Can we find air all around us?

Experience: Part II

PURPOSE: To discover that a candle needs air to burn. To feel the air that exists inside our bodies and to establish that we need air to live.

MATERIALS: Two candles; matches; one clear jar or drinking glass.

Again, discuss the fact that air exists all around us and establish that it would therefore exist in the seemingly empty glass as well as in the entire room. Ask the children which would contain more air—the entire room or the glass. Light the two candles and place the glass over one of the lighted candles. While waiting for the candle flame to go out, discuss which candle has the most air around it. When the flame does go out, ask the children to explain why the other candle remains burning.

Tell the children to test to see if there is any air in their bodies by holding a hand near their noses and mouths and then breathing in order to see if they feel anything escaping from their bodies. Ask whether they need this air to live, just as the candle needs air to burn. If they think they do not need air, have them cover their noses and mouths and hold their breaths for a few seconds. When they gasp for air, ask them why they let air into their bodies.

Suggested Questions

1. Which contains more air—this whole room or this glass?
2. Which candle has more air around it?
3. What happened to the candle that had only the small amount of air in the glass around it?
4. Is there air inside your body? How can you tell?

5. Do you think that your body needs this air to live, just as the candle needs air to burn?
6. What would happen if your body did not get air?
7. What could happen to cause your body not to get enough air? (This is an excellent time to discuss safety rules, such as not playing in old refrigerators.)

Experience: Part III

PURPOSE: To discover that air comes into our bodies through our noses and mouths. To establish that air travels to two balloon-like sacs in the chest that expand as they fill with air and contract as they force the air out.

To discover that air travels from the nose or mouth to the lungs through a pathway or tube called a windpipe or trachea.

MATERIALS: A one-foot section of plastic or rubber tubing (from the hardware store); two balloons; Figure 1-1, "The Heart and Circulatory System."

Ask the children to recall where they let the air into their bodies and where they felt it leaving their bodies in the above experience. Tell them to let a lot of air into their bodies and to watch their chests to see what happens; then tell them to let all the air out and watch their chests to see what happens.

Place two balloons against a child's chest approximately where the lungs would be and explain that there are two balloon-like sacs inside the chest that are called lungs. Tell them to watch as you fill one balloon with air and then let the air out again. At this time, you may wish to introduce the terms *expand* and *contract.* See if the children are now able to suggest why their chests got bigger as they let air into their bodies and why their chests got smaller as they let the air out of their bodies. Tie a knot in the end of the first balloon in order to keep it inflated and repeat this experience using the second balloon. Now use both balloons to

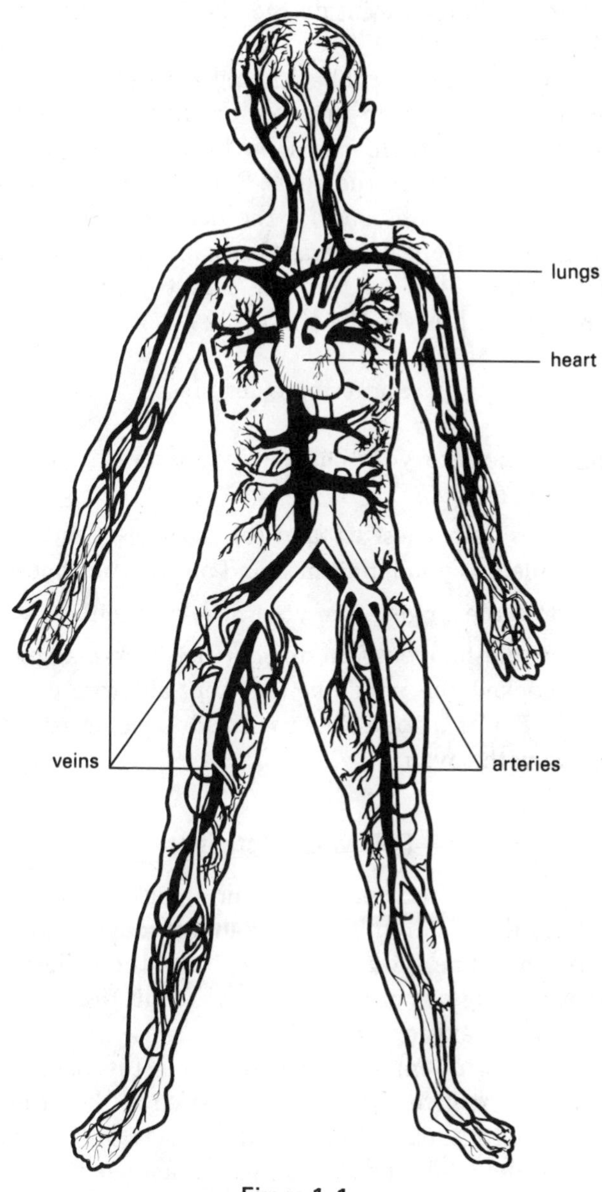

Figure 1-1
The Heart and Circulatory System

show where the lungs are inside the chest and locate the lungs in Figure 1-1.

Show the plastic or rubber tubing and ask if anyone can suggest how the air gets from the nose or mouth to the lungs. Hold the tube near a child in order to show how the windpipe runs from the nose by the mouth to the lungs. Slip the open end of a balloon over one end of the tube and fill the balloon with air by blowing through the tube. Repeat several times, inflating and deflating the balloon in a breathlike rhythm. Be sure to remind the children that there are two lungs in the chest.

Suggested Questions

1. What happens to your chest as you let air in? What happens as you let air out?
2. Why does your chest expand (or get bigger) and contract (or get smaller) as you let air into and out of your body?
3. How does the air get from your nose or mouth to your lungs?
4. What would happen if an object (such as a small piece from a toy) blocked the windpipe and kept air from getting to the lungs? (This is a good time to introduce some safety rules for playing with toys.)

Follow-Up Activities

To let the children "paint with air," place paint on a piece of paper and let the children blow it around through straws to create designs. At this time, you may wish to ask the children whether anything is coming out of the straws, what this thing is called, and how they know it exists.

The activity of inflating and deflating the balloon may be turned into a creative movement experience. Let the children demonstrate with their bodies what happens as the balloon is inflated and deflated and what would happen if the balloon popped.

With older children, you may wish to introduce a discussion of why, at this time, earth is the only location known where humans are able to exist without some special breathing device, why astronauts and divers must wear special suits during their explorations, and why oxygen is provided on airplanes. Let the children design a city that could exist on a planet that has no oxygen in its atmosphere.

Suggested Readings

Branley, Franklin. *Oxygen Keeps You Alive.* New York: Thomas Y. Crowell, 1971.

Golden Press. *My Body—How It Works.* Racine, Wis.: Western Publishing Co., Inc., 1972.

Weart, Edith Lucie. *The Story of Your Respiratory System.* New York: Coward, McCann & Geoghegan, 1964.

THE DIGESTIVE SYSTEM

Information for Adults

Some children have the mistaken idea that after they have swallowed a piece of food, it merely drops down into a large hole in their bodies called the stomach. They have little concept of what happens before the food gets to the stomach or of what the body does with the food once it gets there. The body cannot use food intact in its solid form but, through a process called digestion, converts it to a form that can pass into the blood to nourish all the cells of the body.

Digestion begins in the mouth with the teeth chopping food into fine particles. Here, saliva is added in order to soften the food and begin the digestion of starches. From the mouth, the food is squeezed into the stomach through a tubelike muscle, the esophagus. This tube is separate from the trachea or windpipe, which, except upon rare occasions, automatically closes as one swallows.

The stomach, a pouch or baglike muscle, is rather elastic and can stretch to nearly double its size. In the stomach, food is churned and mixed and undergoes chemical changes. Gastric juice, produced by the stomach, is mixed with the food to form a paste-like liquid called chyme. It takes approximately four hours before food leaves the stomach and is pushed into the small intestine, a tube-shaped muscle more than twenty feet long curled up inside the body. Here, the food is mixed, churned, and squeezed, and more digestive liquids are added to trigger chemical changes. The food is now broken down into minute particles, a form that the blood can use to distribute nutrients to every cell of the body. For the children, the blood may be compared to a milkman carrying the milk (the food that the body can use) to be delivered to every house (cell) along his route.

The part of the food that the body cannot use for nourishment enters the large intestine. Here, some water is reabsorbed, and solid waste material is eliminated from the body as the muscles of the large intestine squeeze it out as feces.

Experience: Part I

PURPOSE: To discover that the teeth chop food and that the mouth adds liquid in order to start the digestive process.

To learn that food travels through a tube to get from the mouth to the stomach.

MATERIALS: Graham or saltine crackers; a dinner knife; a small bowl; a source of water; a section of plastic or rubber tubing (from the hardware store); a diagram illustrating the digestive system (see Figure 1-2).

Begin this experience by letting the children give their ideas of what happens to food after it goes into the mouth. Give each child a cracker to eat and then discuss what is happening as they eat: The teeth are chopping the food, and the mouth is adding a

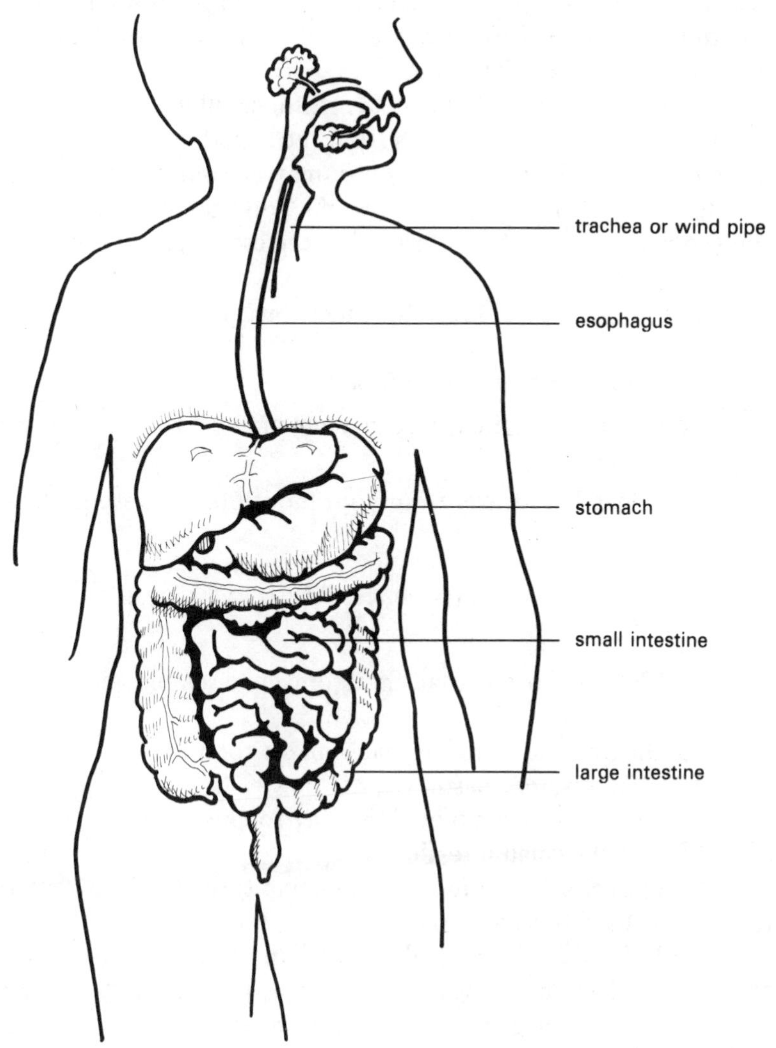

Figure 1-2
The Digestive System

liquid called saliva (they may call this water). In order to illustrate visually what is happening, place a cracker in the bowl and let the children chop it with the knife and add a small amount of water. Save this mixture for Part II.

Show the children the plastic or rubber tubing and ask if they can tell you how the food gets from the mouth to the stomach. If desired, a small object, such as a pea, may be squeezed through the tubing in order to illustrate how the food is squeezed through the esophagus. Help the children find the esophagus in Figure 1-2.

Suggested Questions

1. What do your teeth do to food?
2. Do you feel anything wet in your mouth? What does this do to the cracker?
3. How does food travel from your mouth to your stomach?

Experience: Part II

PURPOSE: To see that the stomach resembles a pouch or bag.

To illustrate how the stomach mixes food and adds liquid in order to form a watery paste.

To observe an example of the way food is squeezed from the stomach into the small intestine.

To illustrate how the small intestine turns food into a form that can be used by the blood.

MATERIALS: The cracker mixture from the previous experience; a diagram illustrating the digestive system (see Figure 1-1), one zip-loc bag; a spoon; a source of water; a tube made from plastic wrap (see Figure 1-3).

Wrap plastic around a cardboard tube at least ten times, leaving one end of the plastic extended about 2 1/2 inches beyond the

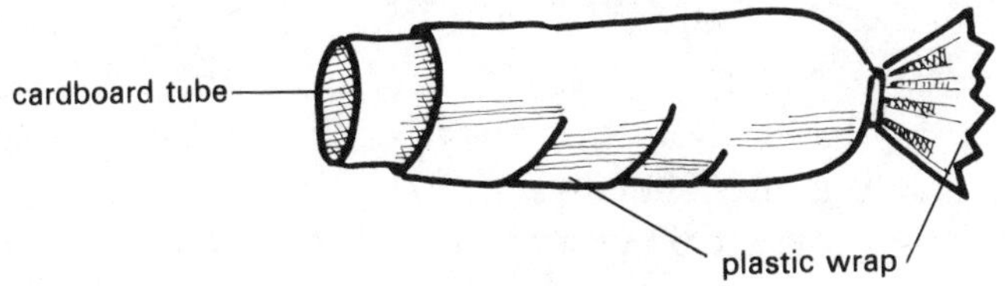

Figure 1-3

tube. Tie off the extended end of the plastic and remove the cardboard tube from the open end of the plastic wrap.

Show the illustration of the stomach muscle in Figure 1-2. Compare the stomach to a pouchlike bag or the the zip-loc bag. Have the children spoon the softened food mixture from Part I into the zip-loc bag. Add more water to the mixture in the bag (enough to form a watery paste when the food is mixed inside the bag). Seal the bag, and then let the children mix the ingredients inside by squeezing. Show the illustration of the small intestine in Figure 1-2 and compare the small intestine to a tube. Help the children squeeze the watery paste from the zip-loc bag into the plastic wrap tubing. Add about 1/2 cup of water, and then let the children squeeze the plastic wrap tube in order to break up the remaining bits of cracker as much as possible. Most of the cracker will dissolve in the liquid. Save this semi-liquid mixture for Part III. Discuss the distribution of food in the blood in terms of the milkman analogy (see Information for Adults). Figure 1-1, "The Heart and Circulatory System," may be used in order to show how the blood distributes food throughout the body.

Suggested Questions

1. What does your stomach do with food? Is there any liquid added in the stomach?
2. How does the food get from the stomach to the tubelike small intestine?
3. What does the small intestine do with food? Is liquid added to the food when it is in the small intestine?

Experience: Part III

PURPOSE: To illustrate that not all of the food in the digestive tract is dissolved into a liquid.

To illustrate how the body squeezes out the parts of the food that it cannot use.

MATERIALS: The food mixture from the previous experience; a small strainer; a bowl; a tube of toothpaste; a diagram illustrating the digestive system (see Figure 1-2).

Let the children pour the liquid food mixture from Part II through the strainer into the bowl in order to collect the solid material that remains. Compare this to the solid waste material that the body cannot use, but explain that the body does not use a strainer. Help the children find the large intestine in the diagram of the digestive system (Figure 1-2), and then see if they are able to suggest how the wastes leave the body. Squeezing toothpaste from the tube will illustrate how the body squeezes out wastes.

Suggested Questions

1. Are our bodies able to use every part of the food that we eat?
2. What happens to the part of the food that the body cannot use in the blood?
3. How does the body eliminate (get rid of) the part of the food that it cannot use?

Follow-Up Activities

This is a good time to introduce the four basic food groups; (1) milk and milk products, (2) meat and eggs, (3) vegetables and fruits, (4) bread and cereals, and to teach nutrition. The body must have a variety of food from all four of the food groups daily in order to get what it needs for good health. Illustrate each of the four basic food groups with pictures, and then play a "what's missing game," showing three foods, one each from three of the four food groups. Ask the children to tell which group is missing.

Let the children plan menus, using the pictures of the four basic food groups. Discuss what empty calorie foods are (those with no nutritive value, such as soda pop and candy) and why they should not be included in the menus. With older children, you may wish to introduce a discussion of why people eat foods that are not nutritious and why there are so many "unhealthy foods" in the grocery store.

Suggested Readings

Krauss, Ruth. *The Growing Story*. New York: Harper & Row, Pub., 1947.

Showers, Paul. *What Happens to a Hamburger*. New York: Thomas Y. Crowell, 1970.

THE HEART AND CIRCULATORY SYSTEM

Information for Adults

The heart is a pumping machine that circulates blood throughout the body. It is approximately the size of a clenched fist, and it is actually a strong muscle made up of four chambers (the right and left atrium and the right and left ventricle) with doors, called valves, between each chamber. As the heart muscle tightens and relaxes, the valves open and close, creating the sound of the heartbeat.

Blood travels from the heart to the lungs for oxygen, back to the heart, and then throughout the body in tubes called arteries. This blood is bright red as it distributes oxygen from the air and digested food particles to all parts of the body. When the blood is depleted of the supply of oxygen and food, tubes called veins return it to the heart for a fresh supply. Blood will again circulate through the heart to the lungs in order to release waste products and to get a new supply of oxygen; then it returns to the heart to be recirculated in the body (see Figure 1-4). Blood in the veins will appear blue in color because it lacks oxygen. Should a vein be cut, the blood flowing out will appear red as it comes in contact with the air.

For the benefit of the children, the circulation of the blood from the heart throughout the body may be compared to a milk

Figure 1-4

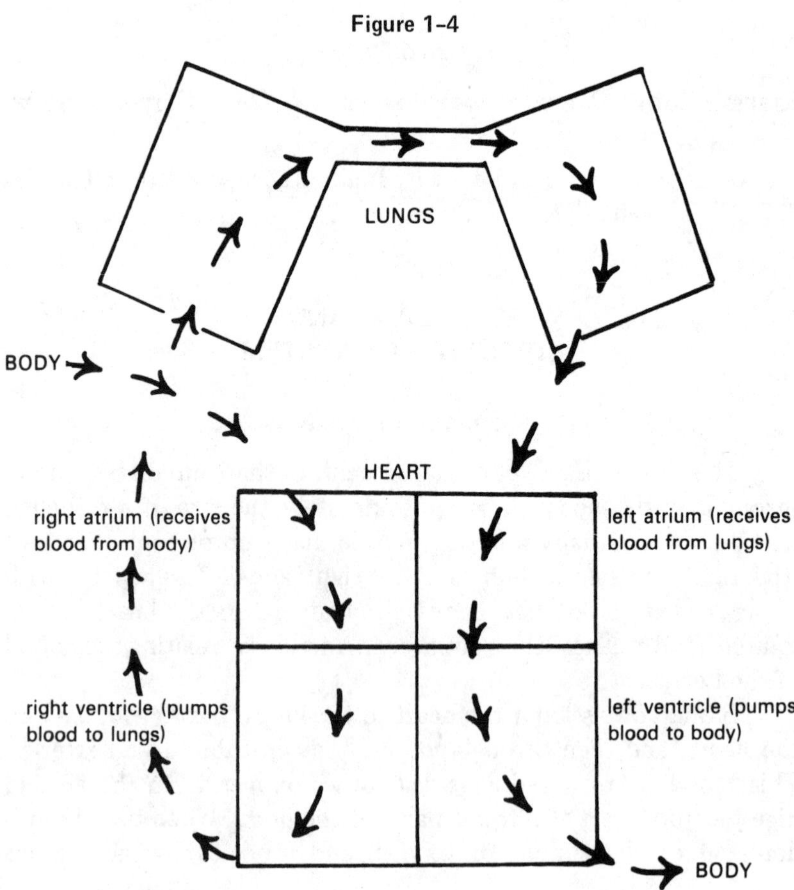

truck making deliveries along its appointed route and then returning to the dairy along a different route in order to pick up more supplies when the truck is empty.

Experience: Part I

PURPOSE: To hear the sound the heart makes as it tightens and relaxes.

To discover that the heart beats faster after exercise.

MATERIALS: A diagram of the heart (see Figures 1-1 and 1-5); a bulb syringe, such as an infant's nasal syringe (from the drugstore).

Have each child place one ear against another child's chest and listen to the sound inside. Let the children mimic the sound of

Figure 1-5
The Heart

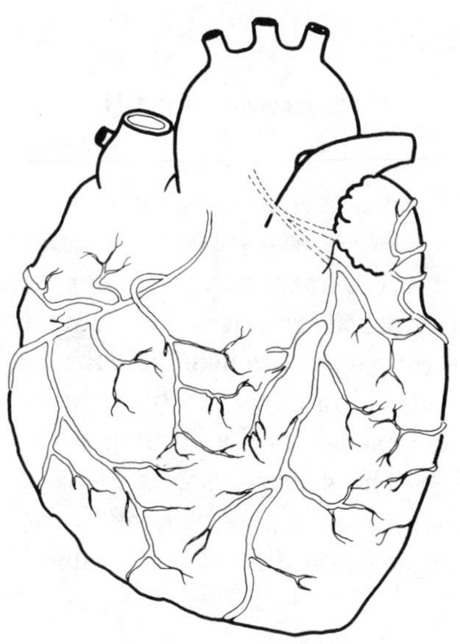

the heartbeat—lub-dup, bump-bump, and so on. Show the diagram of the heart in Figure 1-1, and then ask the children what is making the sound inside their chests. A child's heart is approximately the size of that in the illustration in Figure 1-5.

Have the children jump up and down in place for approximately thirty seconds, and then have each one listen to another's heart after exercise to see if it beats faster.

Ask the children to think of the syringe as a heart, and then have them listen to the sound it makes as they squeeze it in and out. Compare the syringe to the heart muscle tightening and relaxing and compare the sound it makes to that of a heartbeat.

Suggested Questions

1. What is making the sound inside your chest?
2. Does the heart beat faster after exercise?
3. Does the syringe make a sound when it is squeezed in and out?
4. What do you think causes your heart to make a sound?

Experience: Part II

PURPOSE: To discover that the heart is a muscle that pumps blood.

To discover that the heart pumps harder after exercise.

To learn that blood travels away from the heart in tubes called arteries and returns to the heart in tubes called veins.

To observe that blood travels throughout the body and to learn that this blood carries food and oxygen.

MATERIALS: The syringe used in Part I; clear plastic tubing from the hardware store (the pipe or neck of the syringe should fit into the clear tubing); a bowl of water colored red with food coloring.

Ask the children to think of the red water as blood. Squeeze water into the syringe; then squeeze the syringe in order to force the water out so that the children may see what the syringe "heart" does with the blood. Use Figure 1-5 to explain that the heart does not look like a syringe; explain that only the pumping action of the two is being compared. Have the children place their hands on their chests to feel their hearts pumping blood, both before and after exercise.

Show the plastic tubing that represents an artery, suggesting how blood travels from the heart throughout the body. Fit the pipe of the syringe into the tubing, and let the children squeeze the syringe to pump "blood" through the "artery." Figure 1-1 may be used to help show the circulation of the blood throughout the body.

Have the children look for red arteries on their eyelids and behind their ears and look for blue veins at their wrists and inside their forearms. Help them find the pulse at their wrists so that they can feel blood pumping through an artery.

Let the children shine a flashlight through their hands or the inside of their cheeks in order to see the blood inside their bodies. Ask them to give their ideas on why blood is needed in their bodies. Discuss the distribution of oxygen and food in terms of the milk truck analogy.

Suggested Questions

1. What does the syringe do with the red water?
2. What do you think the heart does with blood?
3. Does your heart pump harder after exercise?
4. In what does blood travel throughout the body?
5. What would happen if your heart stopped pumping?

Follow-Up Activities

If possible, get a real stethoscope and demonstrate its use.

Let the children make their own stethoscopes by attaching a piece of rubber or plastic tubing to the narrow end of a small funnel or to the top of a small bleach bottle that has been cut off to a funnel shape. A heartbeat may be heard with the homemade stethoscope when the open end of the funnel is placed against the chest and the loose end of the tubing is placed next to one ear.

Set up a doctor's office for dramatic play. The children will know that the doctor listens to their hearts when they are examined.

Suggested Readings

Clure, Beth, and Helen Rumsey. *Me.* Glendale, Calif.: Bowman Publishing Co., 1963.

Golden Press. *My Body—How It Works.* Racine Wis.: Western Publishing Co., Inc., 1972.

Showers, Paul. *A Drop of Blood.* New York: Thomas Y. Crowell, 1967.

BONES: THE SKELETON

Information for Adults

Bones form the framework of the body, giving form, support, and protection for delicate inner organs. All together, the bones make up the skeleton, which, for the children, may be compared to the framework of a house—the bones representing the individual boards and the skeleton representing the complete frame that gives the house support and shape.

There are 206 bones in an adult's body, and there are more in the body of a child, because certain small bones fuse in the growth process to form one larger bone. An example of this fusing is the skull. Some children will be familiar with the "soft spot" or fontanel on an infant's head. Many bones, such as those of the skull, offer protection for delicate inner organs. The skull offers protection for the brain; the ribcage and breastbone protect the lungs and heart; the vertebrae protect the spinal cord; and the pelvis offers some protection for organs in the lower part of the abdomen.

The places where bones of the body join together are called joints, and these are of several varieties. For the children, joints may be grouped into three main categories: immovable joints, such as those in the skull; slightly movable joints, such as those in the spine; and freely movable joints, such as those in the arms and legs. Freely movable joints of interest to children are hinge joints and ball-and-socket joints. Hinge joints are found in the elbows and allow for movement in only one direction; ball-and-socket joints are found in the hip and shoulder and allow much more freedom of movement, since they consist of one bone moving around freely in the hollowed-out space of another bone. Joints in the wrists and ankles are a combination of hinge and ball-and-socket joints.

Bones, which grow in length as well as in thickness, contain living tissue to allow for growth and minerals, which give strength. Minerals also cause bones to be more brittle and therefore to be more easily broken. Children's bones, which contain more organic matter and a lower content of minerals than those of adults, are more flexible and less easily broken. Bones, even those of an adult, are constantly renewing their cells, which is what gives them the ability to heal when broken. But in order for the healing process to be effective, the bone must be kept stable when mending—hence the need for casts.

Experience: Part I

PURPOSE: To locate bones in various parts of the body.

To observe, in a diagram, that all of the bones of the body form a frame called the skeleton.

MATERIALS: The diagram of a skeleton, Figure 1-6.

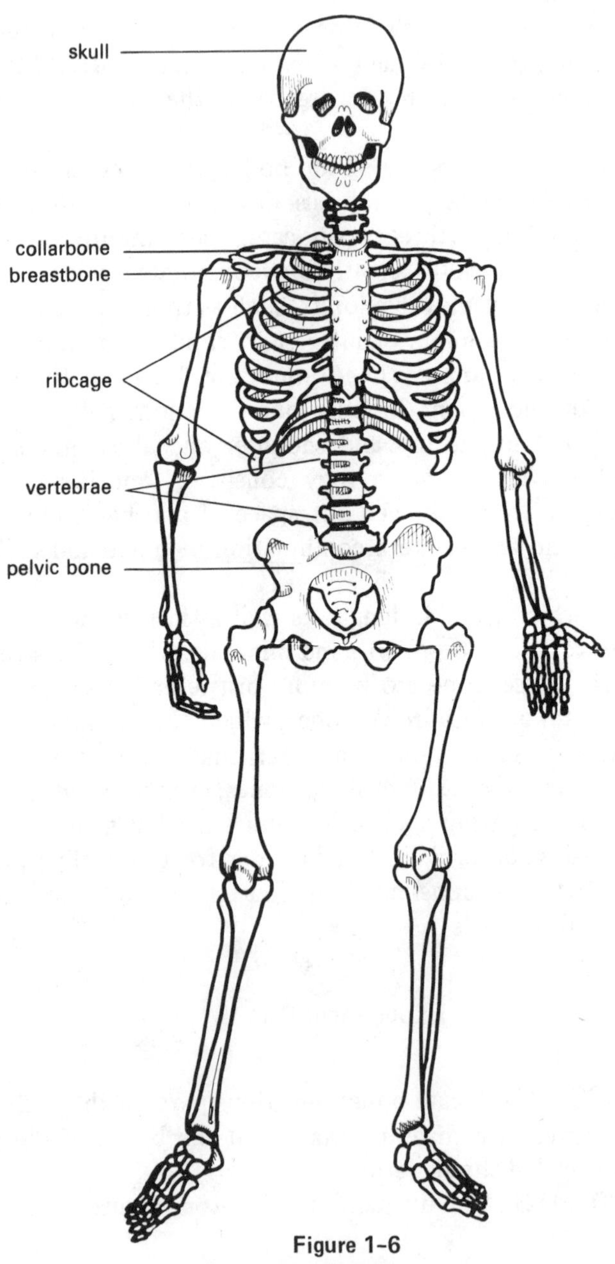

Figure 1-6
The Skeleton

Ask the children whether or not they have bones in their bodies and if they can show you where some of these bones are. Suggest that they feel their bones in the following places: skull, jaw, cheek, neck, collar, breast, ribs, back, pelvis, arms, hands, legs, and feet. Show the picture of the skeleton, and then discuss the analogy of the frame of a house.

Suggested Questions

1. Where are there bones in your body?
2. How many bones do you think you have in your body?
3. How will your bones get to be as large as your mother's or father's bones?

Experience: Part IIA

PURPOSE: To compare the skeleton to a frame that gives shape to and support for the body.

MATERIALS: A scarf; two small frames made of tinker toys or other child's building materials. The frame should show contrast in size and shape (see Figure 1-7).

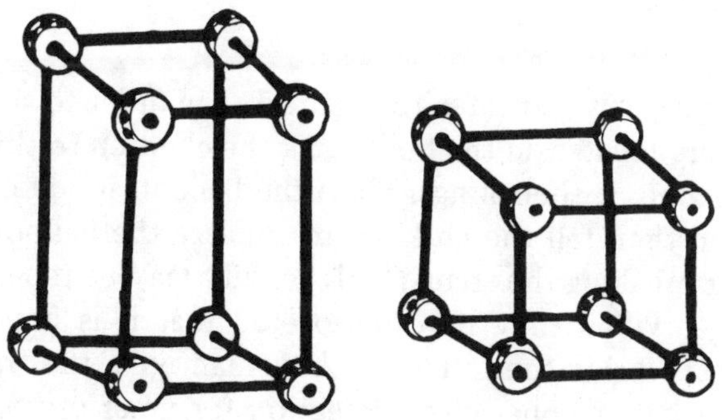

Figure 1-7

Show the children the frames and ask them to tell you how the frames are different. Place the scarf on a table and ask whether it falls into a definite shape or whether it just falls into a pile with no special shape. Ask a child to stretch the scarf tightly over each frame. Now discuss whether the scarf has a definite shape and whether that shape changes with the shape of the frame supporting it. Ask the children to think of the frame as a skeleton and the scarf as skin, and compare with them the function of the skeleton in the body to the function of the frames.

Suggested Questions

1. How would your body look if there were no skeleton inside?
2. How can the shape or size of your skeleton change the shape or size of your body?

Experience: Part IIB

PURPOSE: To discover that a skeleton offers protection for delicate inner organs.

MATERIALS: The two small frames used in the previous experience; a small but slightly inflated balloon to represent a lung and/or a syringe to represent the heart; a diagram of a skeleton (see Figure 1-6).

You may wish to refer to Figure 1-1 in order to show where the lungs are located in the body. Ask the children to think of the balloon as one of their lungs. Place the balloon inside one of the frames, and then tell the children to squeeze the balloon without bending or breaking the frame. This activity may be repeated using the syringe. When they have discovered that it is impossible to harm objects inside the frames without damaging the frames, discuss the fact that bones give protection for inner organs the way the frames offer protection for the balloon and the syringe. In the diagram of the skeleton, find some places where bones are protecting inner organs (the skull, ribs, etc.). Some children like to think of the ribs as a cage and and the skull as a helmet. Let the children feel some of the protective bones in their bodies.

Suggested Questions

1. What do the frames do for the balloon or syringe inside them?
2. If your skeleton is like a frame for your body, what could it do for the parts inside it? For instance, what can it do for the heart, the lungs, or the brain?
3. What would happen if the bones protecting one of these parts were broken?
4. Can you show on your own body where there are bones protecting inner parts?

Experience: Part III

PURPOSE: To learn that the place where two bones are joined together is called a joint.

To learn that different joints have different names.

To discover that some joints allow for much movement and some joints allow for only a little movement.

MATERIALS: A door hinge (from the hardware store); a ball-and-socket connection, such as that on a desk pen set, a shower head, or a swivel lamp; several child's snap-together beads; a diagram of a skeleton (see Figure 1–6).

Ask the children to find some places where their skeletons can move or bend. If they do not know the names for these places, tell them that they are called joints. Discuss what is meant by immovable joints, and then help the children to find these in the diagram of the skeleton. Discuss what is meant by slightly movable joints. Show the children the snap-together beads, and then ask if they can find some joints in their bodies that can move like these beads. Discuss what is meant by freely movable joints. Show the hinge and the ball-and-socket connections, and then ask the children to find some joints in their bodies that are able to move like these.

Suggested Questions

1. Where can your body move or bend?
2. Where are there joints in your body that cannot move at all?
3. Where in your body are there joints that move like snap-together beads, like a hinge, like a ball-and-socket?
4. Do all movable joints move in the same way?
5. What would your body look like if you had no skeleton? How would you walk, sit, or eat?

Follow-Up Activities

Compare the frames of various children: small, large, short, and tall. Ask the children how they can tell without seeing someone's skeleton whether that person has a frame that is large or small, short or tall.

In order to reinforce the idea that the skeleton gives shape to the body, show pictures of the skeletons of various animals, and then see if the children are able to identify the animals just by looking at their skeletons.

Let the children draw pictures of what they think they would look like if they had no skeletons.

Boil a chicken and remove all the bones. Let the children examine the bones and try to put them back together again to form the chicken's skeleton.

If the children seem particularly interested in the study of bones, arrange a field trip to a hospital, doctor's office, or museum where they may see a human skeleton. Hospitals and doctors may also be able to provide x-rays of bones, which make interesting study.

Suggested Readings

Colognese, Don. *The Skeleton Inside You.* New York: Thomas Y. Crowell, 1971.

Weart, Edith Lucie. *The Story of Your Bones.* New York: Coward, McCann & Geoghegan, 1966.

MUSCLES

Information for Adults

Muscles might be called the "movers" of the body. Lifting an arm, blinking an eye, swallowing, breathing, digesting food, and the beating of the heart are all movements of the body that require muscles. Every bone in the body that can move has a muscle, or several muscles, attached to it. These skeletal muscles are attached to the bones by tissues called tendons.

There are approximately 600 muscles in the body, all of which move on command from the brain. Messages travel from the brain through nerve cells in the muscles. Muscles attached to bones are called voluntary muscles and act in response to a thought in the brain. Muscles like those of the heart or digestive tract are involuntary and function without conscious thought, but they are still controlled through nerve response.

All muscles, voluntary or involuntary, use fuel derived from food taken into the body. For the children, this use of fuel may be compared to the use of gasoline by a car. The car cannot move (without being pushed or pulled by an outside force) unless it has fuel. The car, like a muscle, must also have messages (from the driver rather than a brain) in order to direct its movement.

Muscles are able to move a bone only by pulling it. When a

muscle tenses or shortens, it pulls the bone to which it is attached into position; when it relaxes or lengthens, the bone returns to its original position. Many of the voluntary muscles work in pairs, essentially taking turns relaxing and contracting. Muscles change shape as they contract and relax, appearing to bulge as they contract and then returning to their former shape as they relax. Muscles that are used a great deal become larger and will bulge more when contracted.

The activity of a muscle may be illustrated by a model of the arm (made from three cardboard tubes taken from the center of a roll of plastic wrap or wax paper, some string, and a small balloon). Connect the cardboard tubes with string, as shown in Figure 1-8. Then attach a slightly inflated balloon to the tubes as shown. As the tubes are pulled outward or brought inward, the balloon demonstrates the stretching and relaxing of a muscle in the arm.

Figure 1-8
Model Arm

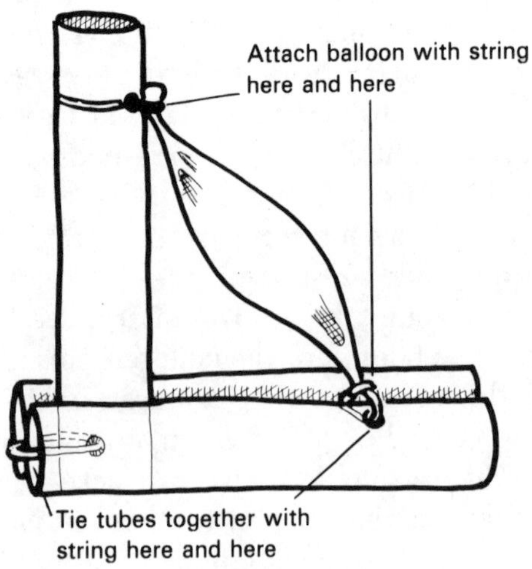

Experience: Part I

PURPOSE: To feel various muscles in the body.

To discover that muscles harden or tighten when working and soften or relax when not working.

MATERIALS: None.

Ask the children whether they have muscles in their bodies, and then suggest that they show you some of these muscles. Have them feel as many muscles of the body as possible. Suggest that they feel various muscles both when they are working and when they are not working. Have them feel cheek muscles when smiling, forehead muscles when frowning, jaw muscles when opening and closing the mouth, muscles in the throat when swallowing, muscles in the arm when lifting an object, and muscles in the legs when walking or jumping.

Discuss the movement of muscles and the necessity of fuel for muscles in terms of the car analogy (see "Information for Adults").

Suggested Questions

1. Where in your body are you able to feel muscles?
2. How do your muscles feel when they are working?
3. How do your muscles feel when they are not working?
4. What kinds of foods do you think would make the best fuel to help your muscles work?

Experience: Part II

PURPOSE: To find out that as muscles tighten and relax (lengthen and shorten) they move parts of the body.

To observe an example of what muscles look like when working.

To establish that muscles are attached to bones by tendons.

MATERIALS: A strong rubber band; the arm model illustrated in Figure 1-8.

Have the children feel the muscle in the top of the upper arm (the bicep) tighten as they raise the arm to a 45° angle. Feel the muscle on the under side of the arm (the tricep) relax at this time. Now tell the children to hold the arm down straight next to the body in order to feel the top muscle relax and the muscle on the under side tighten.

Place the palms of your hands together with a strong rubber band encircling your fingers. Pull your hands as far apart as possible while the children feel the rubber band tighten and watch it stretch. Return your hands to their original position with palms together while the children feel the rubber band relax and watch it shorten. The tightening and relaxing, lengthening and shortening of the rubber band may be compared to the action of a muscle in the body.

Allow the children to operate the model arm in order to see the change in appearance of the "balloon muscle"—bulging when the arm is bent, flattening when the arm is extended.

Discuss how muscles are attached to bones, and then let the children feel the cordlike tendons in their hands and heels.

Suggested Questions

1. What happens to the length of the rubber band as my hands are pulled apart?
2. How does the rubber band feel when my hands are pulled apart?
3. If a muscle works somewhat like a rubber band, what do you think happens to the length of a muscle when it is working? What happens when it is not working?

4. After seeing the model arm, can you tell me what a muscle looks like when it is working? What does it look like when it is not working?
5. When we say that a muscle is working, what is it doing to some part of the body?
6. Why do your bones need muscles attached to them?

Experience: Part III

PURPOSE: To observe that muscles serve a secondary function—filling out the skeleton.

MATERIALS: A scarf; the leg bone of a chicken; a small, slightly inflated balloon.

Tell the children to think of the scarf as skin. Cover the bone with the scarf so that the children may observe how skinny the chicken's leg looks when there is only bone and skin. See if they are able to tell you what is missing. Now tell them to think of the balloon as muscle, which you will place next to the bone. Cover both with the scarf, and then discuss the secondary function of muscles—filling out the skeleton and giving form to the body.

Children will probably think of body form or fullness in terms of fat. Discuss that people who use their muscles a great deal may have forms that are filled out with large muscles. People who use their muscles very little may have forms filled out with more fat then muscle. Body type as well as muscle development helps to determine body form. This activity and discussion may lead the children to the realization that the part of the chicken (or any meat) that they eat is the muscle and the fat (and, in some cases, the skin).

Suggested Questions

1. How would your body look if it had no muscles?

2. What do muscles do for the body other than move the bones?
3. How is the way muscles fill out the body different from the way fat fills out the body? Do muscles look or feel different from fat?

Follow-Up Activities

Discuss, in terms of strength and appearance, what happens to muscles that work hard all the time. Talk about some activities or occupations in which large muscles are developed in certain parts of the body—football player's legs, dancer's legs, weight lifter's arms, ice skater's legs, and so on.

Many of the children will be interested in building strong muscles and may enjoy an exercise program designed for this purpose. A study of nutrition may also be introduced very effectively at this time.

Suggested Readings

Clure, Beth and Helen Rumsey. *Me.* Glendale, Calif.: Bowman Publishing Co., 1968.

Elgin, Kathleen. *The Human Body: The Muscles.* New York: Franklin Watts, Inc., 1973.

Golden Press. *My Body—How It Works.* Racine, Wis.: Western Publishing Co., Inc., 1972.

THE SKIN

Information for Adults

The skin can be compared to a large rubber glove covering the body: holding in fluids to keep inner parts moist, protecting delicate underlayers, and regulating body temperature. Like a rubber glove, the skin will stretch in order to allow for movement or for weight gain. Do not let the children confuse stretching with growing. The skin, unlike a rubber glove, can grow as the body grows—not just stretching to allow for growth, but producing new cells.

Skin comes not in tones of black and white but in variations of brown tones. Two chemicals give skin its color: melanine, responsible for brown coloring; and carotene, responsible for yellow coloring. Various combinations of these chemicals are responsible for the wide variety of skin colors. With the children, this may be discussed in terms of pigment. The more pigment skin contains, the darker it will be—and vice versa. Freckles on the skin are spots with a high concentration of melanine, or pigment. When skin tans, it produces more pigment, and people who have more melanine, or pigment, in their skins will tan more quickly, can become darker, and are less likely to be sensitive to the sun.

Experience: Part I

PURPOSE: To compare the skin that covers the body to a glove that covers a hand.

To discover that skin helps to regulate body temperature.

To learn that skin helps to keep inner body parts moist.

MATERIALS: Disposable rubber gloves (from a drugstore or a beauty supply shop).

Have each child cover one hand with a rubber glove. Compare the glove covering the hand to the skin covering the body. Ask the children to stretch the rubber glove and then to find places where their skin stretches. Let them give their ideas on why skin needs to stretch. Discuss the difference between stretching and growing.

After a few moments, the hand with the glove should feel warmer than the bare hand. (The children may need to wave both hands in the air in order to notice this difference.) Compare the warmth of the glove covering the hand to the warmth of the skin covering the body.

Ask the children to remove the glove and see if that hand feels moist or wet. Have them wave that hand in the air in order to evaporate the moisture quickly. Ask if that hand feels cooler when the moisture goes away. Compare this to perspiration evaporating from the skin to cool the body.

Discuss how the moisture that the skin holds in the body helps to keep inner organs moist or helps to keep them from drying out.

Suggested Questions

1. In what ways is skin like a rubber glove?
2. Do you think skin could help keep moisture in our bodies the way a rubber glove keeps moisture on our hand?
3. What would happen to our inner body parts if we didn't have skin?
4. How can moisture on our skin change the temperature of our bodies?
5. Can we lose skin from our bodies? How? Can you feel the loss of skin when you scrape your knee or elbow?

Experience: Part II

PURPOSE: To observe that everyone's skin is different.

To discover that the color of a person's skin is determined by the amount of pigment it contains.

MATERIALS: Three cups or small bowls of water; one piece each of black, white, and brown construction paper; eye dropper; concentrated brown tempera paint.

Ask the children to name the different colors of skin. They will probably say that skin is black or white. Accept this answer now. Then show them the pieces of black and white construction paper and ask them to find someone in the room with skin that is either black or white, like the paper. Now show the brown construction paper, suggesting that skin comes in tone of brown.

With the eye dropper, add drops of brown tempera paint to the cups of water. Add one drop of paint to the first cup; add five drops of paint to the second cup; add ten drops of paint to the third cup, mixing well each time. Ask the children to compare the variations in color in terms of light and dark. Discuss the reason for these variations. Compare the paint in the water to the pigment in the skin.

Suggested Questions

1. Is skin more brown than black and white?
2. Which water is darker brown? Which water has the most paint in it?
3. Which water is lighter brown? Which water has the least paint in it?
4. Why is light skin (what we call white skin) lighter then dark skin (what we call black skin)?
5. From where do we get the color of our skin?
6. Is the color of a person's skin important? Does it change or affect the person's personality in any way?

Follow-Up Activities

Compare skin differences, such as freckles or no freckles, tan or no tan.

Discuss color changes in the skin when it is exposed to sunlight.

Differences in the texture or patterns of skin may be compared by having each child make fingerprints and handprints with tempera paint and white paper.

Suggested Readings

Evans, Eva Knox. *All About Us.* New York: Golden Press, 1965.

Showers, Paul. *Your Skin and Mine.* New York: Thomas Y. Crowell, 1967.

SOUND AND THE VOICE

Information for Adults

Sound begins with a vibrating body—that is, one that is moving rapidly back and forth or up and down. Vibrations produce sound as they set up a type of energy that creates waves, called sound waves. Sound waves travel in a spherical pattern, moving outward from the vibrating body. The way in which sound waves travel might be compared to the way waves or ripples spread outward in a spherical pattern after an object has been dropped into a body of still water.

When sound waves meet a solid or massive object (a wall or a thick wooded area, for example), the waves may be reflected off the object much the way a ball may bounce off a wall. The waves are then directed back toward their original source, creating an effect known as an echo. The creation of an echo may be demonstrated, in part, by dropping a small object into the center of a pan of water. The waves that will be created will travel outward from the object, meet the side of the pan, and then be reflected back toward the center of the pan. For the children, echoes may be termed "bounced sounds."

The sound of the voice begins in the larynx or voice box, a tubelike structure in the throat at the top of the trachea, or windpipe. Stretched across the larynx are folds of muscle called vocal cords. It is the vocal cords that are mainly responsible for creating the sound of the voice. For the children, a loose comparison may be made between the vocal cords and rubber bands: Both may vibrate, and both may stretch. As air is forced across the vocal cords, vibrations that create sound are set up in the throat. The vocal cords also act as a valve that can decrease or increase the amount of air going through the throat. When one takes a breath, the vocal cords open in order to allow air to pass without vibration.

Sounds created by the vocal cords can change according to the force with which the air from the chest vibrates the cords, changing the loudness or softness of the sound. Signals from the brain can change the tension of the vocal cords, creating sounds of higher or lower frequency or pitch. The mass of the vocal cords also contributes to pitch; the more massive vocal cords of most men create a lower pitched sound than the less massive vocal cords of most women and children.

Changes in the quality of a sound also take place in the vocal cavity (the throat, mouth, and nasal cavities), where air begins to vibrate. Changes in sound to create speech are produced with the tongue, the lips, and the teeth.

Experience: Part I

PURPOSE: To observe and describe what happens when an object vibrates.

To discover that vibrations create sound.

To discover that sound changes as the quality of the vibrations change.

MATERIALS: A plastic ruler; a toy drum; a few grains of uncooked rice.

Place the plastic ruler on the edge of a table top with most of the length of the ruler extended over the edge. Hold the ruler firmly on the table while a child hits the free end in order to set the ruler into a vibrating motion. Ask the children to describe what they observe. Tell them that this up and down movement is called vibration.

Place a few grains of rice on the top of a toy drum, and then have the children tap the top of the drum in order to see the rice bounce up and down with the vibrations. Let the children describe what they have observed, and then see if they are able to identify this movement of the drum as vibrations. Ask whether a sound was made as the top of the drum vibrated.

Again place the ruler on the edge of the table with most of the length extended over the edge. Ask whether the ruler is making a sound. Have a child vibrate the ruler in order to discover that a sound is made. Repeat the activity with seven inches (and then with three inches) of the ruler extended over the edge of the table. In each case, ask what is different (Is the ruler shorter? Does it vibrate faster? Does the sound change?).

Suggested Questions

1. What happens to the rice as you tap the top of the drum?
2. What is the top of the drum doing as you tap it? Does it make a sound?
3. What must the ruler do in order to make a sound?
4. How did the sound change as the ruler vibrated faster or slower?
5. In what way did the sound change as there was a longer or shorter part of the ruler vibrating?

Experience: Part II

PURPOSE: To understand that the sound of the voice is made with the vocal cords (muscles that may be compared to rubber bands stretched across the throat).

To discover that vocal cords make a sound as they vibrate.

MATERIALS: A rubber band.

Have the children place one hand against their throats and hum in order to feel the vibrations. Have them give their ideas of what is causing these vibrations. Stretch a large rubber band between your thumbs and compare it to the vocal cords stretched across the larynx. Let the children plunk the rubber band in order to see it vibrate. See if they are able to suggest how the vocal cords could create the sound of the voice.

Have the children try to create a sound with their own vocal cords without allowing any air to pass through the larynx. Suggest that they hold their breaths as they try to talk and that they try not to let any air into or out of their bodies.

Suggested Questions

1. What do you think is causing the vibrations in your throat when you hum?
2. If your vocal cords are something like rubber bands, how do you think they are able to create the sound of your voice?
3. What causes your vocal cords to vibrate?

Experience: Part III

PURPOSE: To discover some ways in which the sound of the voice may be altered.

MATERIALS: Rubber bands of various sizes; a small box.

Stretch a rubber band between your thumbs and have the children plunk it as you change the tension. Ask them to describe what is happening to the sound as the rubber band is stretched and relaxed. Let them experiment with changing the sound or pitch of their voices by stretching and relaxing their vocal cords.

In order to discover how the force with which the air vibrates the vocal cords changes the quality of the sound, stretch the rubber band tightly. Have the children plunk it gently and then vigorously and ask them to listen for changes in the sound. Tell them to say "ah"—first, while letting the air pass very gently through their throats; then, while vigorously forcing the air through their throats.

In order to compare how the size or mass of the vocal cords may change the quality of the sound of the voice, stretch rubber bands of various sizes across a small box, and plunk each band to listen for differences in the sound.

In order to discover how the tongue, lips, and teeth change sounds to help create speech, let the children experiment with the following activities: Hold the tongue between the fingers and try to pronounce words that begin with the letter "d," "b," "s," and "p." Attempt to repeat a phrase, such as "Peter, Peter, pumpkin eater" or "Mickey's magic monkey," without allowing the lips to come together. Try to pronounce the word "oh" without forming the mouth into an "o" shape. Ask a child who has front teeth missing to repeat a phrase like "She sells sea shells."

Suggested Questions

1. How does the sound that the rubber band makes change as the rubber band is stretched or relaxed?
2. How might the sound of the voice be made higher or lower?
3. How does sound change when something is vibrated gently and when it is vibrated very hard?
4. How can you make your vocal cords vibrate very hard?
5. What might be a reason why most men's voices are different from most women's voices?
6. What parts of our mouth help us to form words?

Follow-Up Activities

Let the children feel the vibrations of various sound-making devices, such as a ticking clock; a record player playing a record loudly; a small appliance, such as a blender or an electric mixer; a radio playing loudly; and the like. In each case, discuss what is causing the object to make a sound.

Make musical instruments that demonstrate how vibrations make sounds. A guitar may be made by stretching rubber bands around an empty box. Create a sort of harmonica by folding a piece of wax paper over the teeth of a comb. Press your lips gently against the paper and hum in order to vibrate the paper and make a sound.

The adult may wish to continue the study of sound and how it travels by demonstrating an example of the creation of an echo. Set up a pan of water in which you create waves that may be compared to sound waves. The waves in the water will bounce off the sides of the pan in much the same way that sound waves are reflected off a solid object. Note: a large pan of water (at least 9″ × 13″) is desirable for this activity.

Suggested Readings

Branley, Franklin. *High Sounds, Low Sounds.* New York: Thomas Y. Crowell, 1967.

Showers, Paul. *How You Talk.* New York: Thomas Y. Crowell, 1966.

——. *The Listening Walk.* New York: Thomas Y. Crowell, 1961.

White, Laurence B. Jr. *Investigating Science With Rubber Bands.* Reading Mass.: Addison-Wesley, 1969.

SOUND AND THE EARS

Information for Adults

An object may vibrate and send sound waves traveling through the air, but if the sound is to be heard, there must be some sort of listening device. The human ear is a very effective listening device made up of three parts: the outer ear, the middle ear, and the inner ear.

The outer ear begins as an opening, surrounded by cartilage, on either side of the head. It is this cartilage that helps to direct sound waves into the ear. Animals are able to use this part of the ear very effectively, turning the ears to pick up sounds more directly.

The outer ear continues as a small passageway, called the ear canal. The canal, about one inch in length, acts as a sort of amplifier, increasing the pressure of the sound wave in the ear. The ear canal ends with a piece of membranous skin, called the eardrum, stretched across it. Sound waves entering the ear vibrate the eardrum at the same frequency as the sound wave.

Beyond the eardrum is the middle ear, an air-filled cavity containing three small bones linked together; the bones are named for their appearance—the hammer, the anvil, and the stirrup. Vibrations of the eardrum set these three bones in motion so that they transmit the vibrations to the oval window, a membrane stretched across the entrance to the inner ear (between the stirrup and the cochlea).

Beyond the oval window is the cochlea (kŏk′ lē–á), a spiraled bone chamber filled with fluid. The semicircular canals in the inner ear are not part of the cochlea and have nothing to do with hearing; rather, they help to control balance. Vibrations of the oval window are received by the fluid of the cochlea and are transmitted by hair cells within the cochlea to nerve cells linked together to form the auditory nerve. It is the auditory nerve that carries the sound, or nerve impulses, directly to the brain.

Experience: Part I

PURPOSE: To observe an example of the way sound waves travel.

MATERIALS: A large pan of water; a small object, such as a penny.

Note: The adult may wish to refer to "Information for Adults" in "Sound and the Voice" before beginning this experience. Let a child drop the penny into the center of the pan of water, and then ask the children to describe what they see. Repeat the activity if they did not notice the waves in the water. Ask the children to describe the way the waves travel. Compare these waves, which travel in a circular pattern in the water, away from the penny, to sound waves, which travel in a circular pattern in the air, away from the object that makes the sound.

Suggested Questions

1. What did you see in the water when we dropped the penny in it?
2. In what shape or pattern did the waves travel?
3. Did the waves travel away from or toward the penny?

Experience: Part II

PURPOSE: To discover that sound can travel better through solid objects and through water than it can through the air.

MATERIALS: A wind-up clock; a measuring stick; an aquarium or a gallon jar filled with water.

Hold the clock about thirty-six inches away from a child and tell him or her to listen for the ticking. Ask the child to tell you what the sound is traveling through to get from the watch to his or her ear. Now hold the watch firmly against one end of the measuring stick while the child holds one ear against the other end to listen for the ticking. Ask the child to tell you what the sound traveled through this time. Let the children use the clock to test the way sound travels through other solid objects, such as a metal rod or a table top.

Tap the side of the aquarium or jar of water and have the children listen for the sound. Ask what the sound traveled through to get from the tank to their ears. Now have the children, one at a time, place one ear firmly against the tank or jar as you tap on the opposite side. Ask the children to tell you what the sound traveled through this time to get to their ears.

Suggested Questions

1. Through what did the sound travel better, the air or the measuring stick?
2. What are some solid things that sound travels through?
3. Through what did the sound travel better, the air or the water?

Experience: Part III

PURPOSE: To understand that we hear sounds because of vibrations created inside our ears.

MATERIALS: Toy drum; a few grains of uncooked rice; a diagram of the inside of the ear (see Figure 1-9); three building blocks of the same size; the pan of water used in Part I.

Figure 1-9
The Ear

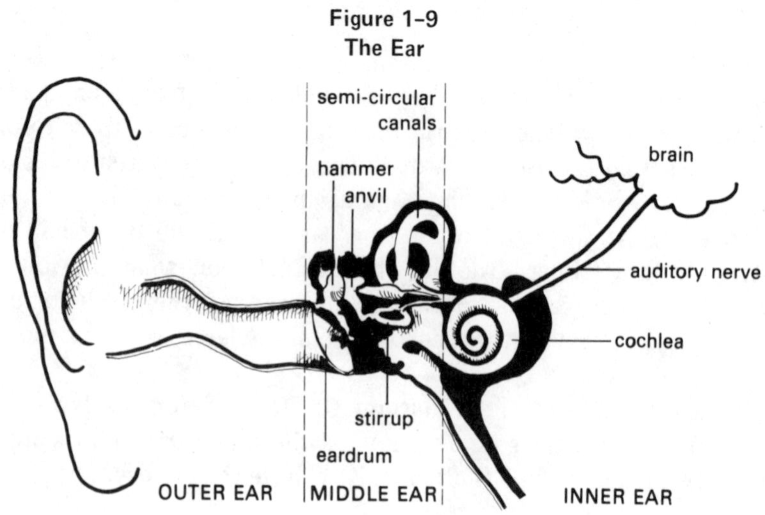

In order to begin to illustrate the series of events involved in hearing a sound, recall Part I of the previous unit, "Sound and the Voice," with the children. Discuss the activity involving the drum and the uncooked rice or present that activity now as follows. Place a few grains of rice on the top of a toy drum and have the children tap the top of the drum in order to see the rice bounce up and down with the vibrations. Let them describe what they have observed, and then see if they are able to identify this movement of the drum as vibrations. Show them the diagram of the ear (Figure 1-9) and compare the eardrum to the top of the toy drum.

In order to show how the eardrum can set up a series of vibrations in the middle ear, line up the three building blocks so that they are standing upright about three inches apart. The purpose of this activity is to set up a chain reaction so that, by bumping the drum, the first block knocks over the second block, and the second block knocks over the third block. Allow the children to experiment with this activity. Then compare the three blocks to the three bones in the middle ear and find the bones in the diagram of the ear.

Discuss the fact that the inner ear contains a liquid in the cochlea, and find this part of the ear in the diagram. Show that the cochlea joins the auditory nerve, which travels to the brain. Ask a child to take one of the blocks and to tap gently on the pan of water. Ask the children to describe what they see in the water. (They may say that the water is wiggling or that it has little waves in it.) Suggest that there are vibrations traveling in the water. See if they are able to suggest how sound waves might travel from the middle ear to the brain.

Suggested Questions

1. How did the toy drum make a sound?
2. When we bumped the drum to make it vibrate, what happened to the blocks?
3. What do you think vibrations of the eardrum could do to the three bones in the ear?

4. What could you see traveling in the water when we bumped the pan?
5. How might sound waves (or vibrations) travel from the three bones in the middle ear to the brain?

Follow-Up Activities

Another way to illustrate how sound travels is to stretch a Slinky-type toy across the top of a table, with one person holding each end. Have one person quickly push his or her hand forward about three inches in order to observe the chain reaction that is set up in the coil. This might be compared to the chain reaction that is set up inside the ear when a sound is received.

Suggested Readings

Branley, Franklin. *Timmy and the Tin Can Telephone.* New York: Thomas Y. Crowell, 1959.

Duff, Maggie. *Johnny and His Drum.* New York: Henry Z. Walck, 1972.

Scarry, Richard. *Great Big Air Book.* New York: Random House, 1972. (see "Orchestra Practice" pp. 56-57).

SEASONS

Information for Adults

This experience is very appropriate for use in the fall or spring. With younger or less intellectually mature children, the adult may wish only to help the children become more aware of the changes that take place with each season and of the fact that seasons keep occurring over and over again in the same sequence. For older children, the experience illustrating the reason for the change of seasons (Part II) should be presented on an individual basis or with a very small group of children.

When discussing seasonal changes or "signs of the seasons," help the children think in terms of changes that are easily observed or easily documented: what is happening with outdoor growing plants (trees, grass, flowers), what temperature and weather changes are taking place (determined by a thermometer as well as by the type of clothes the children are wearing), what is the length of sunlight each day (as measured by the time of sunrise and sunset), what the animals are doing (active or inactive, gathering food or enjoying an available abundance, bearing young, shedding fur or feathers or getting thicker fur or feathers).

Seasonal changes take place because of the earth's relationship to the sun as it makes its approximately 365-day trip around the sun. The earth spins on an axis, an imaginary line running from north to south through its center. As the earth travels around the sun, the axis slants at an angle of about 23 1/2° instead of standing upright. This slant means that at times the northern hemisphere is receiving more direct rays from the sun (summer), whereas the southern hemisphere is receiving less direct rays from the sun (winter). At other times, the reverse is true (see Figure 1-10). The tilt of the earth on its axis is also responsible for changes in the length of the days.

Experience: Part I

PURPOSE: To see pictures representing each season and to discuss the characteristics or signs of each season.

MATERIALS: Pictures representing various signs of all four seasons: Calendars are good sources for pictures, and most of the books listed in Suggested Readings for this topic offer appropriate pictures that may be used.

Begin this experience by discussing in general terms what the signs of the seasons are. See how many signs the children are able to name on their own, stimulating their thoughts with the pictures.

Figure 1-10

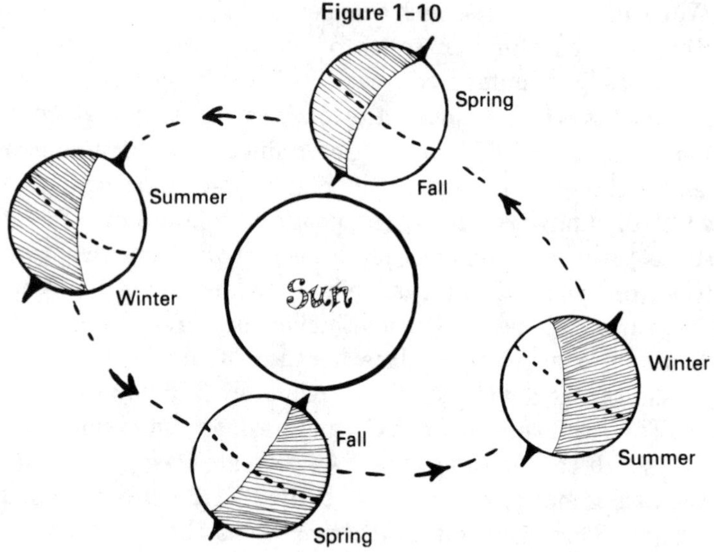

Suggested Questions

1. What do the trees look like in this picture: Are there leaves? What color are they? What season is this?
2. Look at the clothes the people are wearing in this picture. For what kind of weather are they dressed? What season is this?
3. Does this look like a cool or a warm day? What are the people doing (swimming, sledding)? What season is this?
4. What are the animals in this picture doing (migrating, gathering food, building homes, caring for young, etc.)? What season is this?

Experience: Part II

PURPOSE: To illustrate visually that seasons make a full circle, occurring over and over in the same sequence.

MATERIALS: A "seasons wheel," consisting of a large square of poster board with a circle of poster board attached to the center by a paper fastener, so that the circle turns easily. There should be a picture in each of the four corners of the square to represent each of the four seasons, and on the circle there should be a picture of a child (see Figure 1-11).

Discuss which season each of the pictures on the seasons wheel represents; then let the children turn the circle in a clockwise motion. As the circle turns, the child on the circle passes in and out of each season and begins the sequence again at the end of the cycle.

Figure 1-11
Seasons Wheel

Suggested Questions

1. What season is this child in now? How can you tell?
2. What happens when the child has gone through all four seasons?
3. Could the child go through the seasons backwards? Could nature go through the seasons backwards—for instance, could trees lose leaves in the fall if fall came *after* winter when trees are bare of leaves?

Experience: Part III

PURPOSE: To illustrate visually the reason for the change of seasons.

MATERIALS: A globe; two pennies; masking tape; a strong light source, such as a lamp with the shade removed.

Mark the equator on the globe with the masking tape, and then tape one penny to the southern hemisphere of the globe and the other penny to the northern hemisphere of the globe, equidistant from the equator. Hold the globe at approximately a 23 1/2° angle to the light source (the sun). Begin with the position that illustrates spring in your geographic area and move through summer, fall, and winter. The axis should remain at the same orientation to the light at all times as you move the globe around the light source (the sun). Remain at each position representing a season for at least one minute; then let the children feel the pennies. They should be of approximately the same temperature in the spring and fall position, and one will be warmer than the other in the winter and summer position.

Suggested Questions

1. When the weather is hot or very warm, what season are we having?

2. When the weather is cold, what season are we having?
3. If the penny on one half of the globe is hot or very warm, what season are the people in that area having?
4. If the penny on one half of the globe is cold, what season are the people in that area having?

Follow-Up Activities

Because trees represent the seasons so well, let the children make four season trees. Provide each child with four bare tree trunks and let them decorate the trees to represent each season. Shake cotton balls in a paper bag with dry tempera to use for blossoms on a spring tree; twist small pieces of green tissue paper and glue to a tree trunk for a summer tree; sponge-paint leaves in orange, red, and yellow on a fall tree; and spatter white paint over a bare tree to represent snow for a winter tree. It may be simpler to let the children decorate one tree to represent each of the seasons each day for four consecutive days.

Suggested Readings

Birmingham, John. *Seasons.* Indianapolis: Bobbs-Merrill, 1969.

Hoff, Sid. *When Will It Snow?* New York: Harper & Row, Pub., 1971.

Holl, Adelaide. *The Wonderful Tree.* Racine, Wis.: Western Publishing Co., Inc., 1971.

Ichikawa, Satomi. *A Child's Book of Seasons.* New York: Parent's Magazine Press, 1975.

Kane, Henry B. *Four Seasons in the Woods.* New York: Knopf, 1968.

Parker, Bertha Morris. *The Wonders of the Seasons.* Racine, Wis.: Western Publishing Co., Inc., 1974.

Shannon, Terry. *Come Summer, Come Winter.* Chicago: Albert Whitman and Co., 1968.

Tresselt, Alvin. *White Snow Bright Snow.* New York: Lothrop, Lee & Shepard Co., Inc., 1947.

——. *It's Time Now.* New York: Lothrop, Lee & Shepard Co., Inc., 1969.

Zolotow, Charlotte. *Over and Over.* New York: Harper & Row, Pub., 1957.

FALL

Information for Adults

The purpose of this unit is to explore some of the natural phenomena taking place in the fall. It includes experiences dealing with why leaves change color and with how animals prepare for the winter.

Fall color changes in leaves are the result of a combination of occurrences: loss of chlorophyll due to shorter days and chemical changes triggered by cooler weather. For simplicity, with the children, color changes in leaves may be attributed mainly to the loss of chlorophyll. Deciduous trees need much light in order to produce the chlorophyll that gives them their green color. As the season changes to fall and the days grow shorter, the trees have fewer hours of sunlight to produce chlorophyll. Without the production of chlorophyll, leaves begin to lose their green color, and colors of red, orange, yellow, and brown begin to appear. These colors have always been present in the leaves, but they have been covered by the green chlorophyll. With fall, cooler weather also comes, which causes chemical changes to take place, producing other color changes and vividness of color.

The lifestyle of many animals changes greatly as cooler weather comes. This information is intended to give some very general knowledge of the way some familiar animals prepare for winter. Some animals, such as the squirrel, the chipmunk, the beaver, and some mice store food for the winter and thus are very active in the fall. Other animals, such as rabbits, opossums, deer, foxes, and bobcats, make no conscious preparation in the fall but grow thicker fur and stay active all winter. Raccoons, skunks, and bears also grow thicker coats, but they will sleep (not hibernate in the true sense of the word) most of the winter and do not gather food in the fall.

Hibernation is a very deep sleep in which an animal's body functions actually slow down, and the animal does not require food for the entire period of hibernation. Some animals that hibernate are woodchucks, some mice, snakes, salamanders, lizards, and some insects. Frogs and turtles may hibernate buried in the mud at the bottom of a stream or pond.

Squirrels, woodchucks, beavers, muskrats, and bears build or find homes in which to spend the winter. Some bees seal up their hives in order to keep warm. Some birds, butterflies, and fish migrate in the fall, moving to sources of warmth and food. Birds that do not migrate may have difficulty finding food during the coldest months. Fish that do not migrate may move to the bottoms of rivers and ponds where the water is warmer.

Many spiders and insects lay eggs during the fall and then die in the cold, leaving the eggs to hatch in the spring. Some insects find places in the fall in which they are protected from the cold — under leaves or rocks, under the bark of trees, deep under the ground, or in old logs. Some butterflies and moths spend the winter in a pupal stage in a chrysalis (butterflies) or a cocoon (moths). Most adult butterflies and moths die in the winter cold, although some butterflies (monarchs) migrate to warmer climates.

Experience: Part I

PURPOSE: To discover, in part, why leaves change color in the fall.

MATERIALS: Some colored fall leaves; some green leaves from deciduous trees; a mason jar half-filled with a strong household bleach, such as Clorox (weak bleach will not work).

Note: This experience must be presented early in the fall when there are still green fleshy leaves to use. Be sure to test leaves in the bleach before presenting the activity to the children in order to find which leaves bleach most successfully.

Show the children the fall leaves and ask for their ideas of why leaves change color in the fall. Submerge one or two green leaves in the jar of bleach to leave overnight. Screw the lid of the jar on very tightly; then tape the lid securely to the jar so that the children will not be tempted to taste the bleach. The next morning, when the bleach has removed the green color from the leaves, leaving them pale yellow or white, discuss with the children what has happened. Tell them that all of the leaves that change color in the fall have other colors underneath the green but that we cannot see the other colors until the green goes away. You may wish to tell the children that this green coloring is chlorophyll.

Suggested Questions

1. What did the bleach do to the leaves?
2. When the green went away, what colors were there underneath?
3. How, in nature, might the green go away?

Experience: Part II

PURPOSE: To observe that one of the reasons leaves change color in the fall is lack of light.

MATERIALS: A very small deciduous tree growing outdoors; string; a dark-colored plastic bag large enough to cover the tree or a small, green, living outdoor potted plant and a cardboard box large enough to cover the plant).

Find a small deciduous tree growing outdoors, and place a large, dark-colored plastic bag over the limbs and leaves in order to cover all but the lower part of the tree trunk. The bag may be sealed with a string tied around the trunk. A small, outdoor potted plant may be brought indoors and covered with a cardboard box in order to remove any source of light. Discuss with the children that the reason for covering either of the plants in this activity is to remove all the light. It will take two to three weeks (possibly more) for the leaves of these plants to begin to turn yellow. At this time, let the children remove the box or the plastic bag and describe what they see.

With older children, this activity may be carried a step further by letting the children check the time of sunrise and sunset each day in order to record the hours and minutes of sunlight. In one week of doing this, they will begin to see that each day the trees have less time in the sunlight.

Suggested Questions

1. What did we do to the light when we covered the plant or tree?
2. What happened to the leaves when we took the light away?
3. Could the plant or tree make the green color or chlorophyll when there was no light?
4. How could nature take the light away from the plants and trees?

Experience: Part III

PURPOSE: To understand how the fur of some animals and the feathers of some birds help keep them warm in the winter.

MATERIALS: A child's heavy coat; a room thermometer; masking tape.

With the tape, mark the temperature of the room on the thermometer. (It may be necessary to explain to very young children that the red line on the thermometer goes up when the temperature is warmer and down when the temperature is cooler.) Have a child wear the coat for a least two minutes with the thermometer placed next to his or her skin. While waiting for the temperature on the thermometer to rise, discuss with the children that animals wear fur or feathers instead of coats and that these coats or feathers get thicker in the winter. Remove the thermometer and check the temperature. Ask the children which was warmer, the room or the body of the child inside the coat.

Suggested Questions

1. Do animals wear coats to cover them? What do they wear?
2. For younger children, ask, "Did the red line of the thermometer go up when the thermometer was under the coat?"
3. Was the temperature warmer inside the coat than out in the room?
4. Why do animals grow thick fur or feathers in the winter?
5. What happens to the animal's fur or feathers when the weather warms up again? (Have you ever petted a dog or cat and had lots of hair rub off onto your hand?)

Experience: Part IV

PURPOSE: To observe what some fish do in order to stay warm in the winter.

MATERIALS: An aquarium or institutional size jar containing nontropical fish; a plastic container or plastic bag filled with ice cubes.

Turn off the heater in the aquarium and float the container or plastic bag filled with ice cubes on the surface of the water. It may be necessary to let the children feel the water in the aquarium or jar in order to find out that the water near the surface is now colder than the water near the bottom. Observe that the fish move to the bottom of the tank. Ask the children where fish in a pond or lake might find warmer water when the air at the surface of the water is very cold. Discuss migration in terms of another way fish could find warmer water in the winter. Note: Do not let the ice cubes stay in the water for an extended period of time, as extreme cold may harm the fish.

Suggested Questions

1. What happens to the water near the ice cubes?
2. Where in the aquarium or jar do the fish go when the water near the top gets cold?
3. Where do you think the fish in a pond or lake go when the weather makes the water near the surface cold?
4. What else could fish do to find warmer water in the winter?

Experience: Part V

PURPOSE: To discover what some frogs, turtles, or salamanders do when cooler weather comes.

MATERIALS: An institutional size jar or an aquarium half filled with mud and filled to the three-quarter mark with water (let the mud and water sit in the jar overnight, and the next day add ice cubes in order to bring the water level nearly to the top of the jar. Add the ice very carefully so as not to stir up the mud). A frog, turtle, or salamander.

Let the children feel the water in order to establish that it is cold. Ask them to think of the water in the jar as water in a pond when cold weather comes. Place the frog, turtle, or salamander in the cold water. Eventually, it will dig through the mud to the bottom of the jar or aquarium.

Suggested Questions

1. How does the water in the jar or tank feel? During what season would the water in a lake or pond feel this cold?
2. What did the frog do when we put it in the cold water?
3. What can frogs in streams or ponds do when cold weather comes?

Follow-Up Activities

Let the children collect a variety of fall leaves and then categorize them according to color—all the red leaves in one pile, all the orange leaves in another pile, and so on.

Use fall leaves to make "stained glass windows." Cut two sheets of wax paper approximately 8″ × 10″. Let the children arrange the leaves between the wax paper sheets. The adult may press the wax paper with a warm iron to make it cling together and hold the leaves in place. The children may mat their projects with frames made of construction paper. Hang the finished projects in front of a window to allow sunlight to shine through.

In order to help children understand how trees grow new leaves in the spring after losing their leaves in the fall, force buds open from a variety of trees. After at least six weeks of cold weather, remove budded branches and bring them indoors. Keep the branches in water in a warm, sunny place. In three or more weeks (depending upon the type of tree from which the branches were removed), the buds will open to reveal new leaves. Note: The earlier the buds on a tree open in the spring in a natural setting, the earlier they will open indoors.

Suggested Readings

Trees and Plants

Holl, Adelaide. *The Wonderful Tree.* Racine, Wis.: Western Publishing Co., Inc., 1971.

FALL

Kessler, Ethel, and Kessler, Leonard. *All for Fall.* New York: Parent's Magazine Press, 1974.

Petie, Haris. *The Seed the Squirrel Dropped.* Englewood Cliffs, N.J.: Prentice-Hall, 1976.

Tresselt, Alvin. *Autumn Harvest.* New York: Lothrop, Lee and Shepard, 1951.

Animals

Cosgrove, Margaret. *Wintertime for Animals.* New York: Dodd, Mead, 1975.

Hader, Berla, and Hader, Elmer. *The Big Snow.* New York: Macmillan, 1948.

Lubell, Winifred, and Lubell, Cecil. *In a Running Brook.* Chicago: Rand McNally, 1968.

2
WINTER

FROST AND SNOW

Information for Adults

Frost and snow are similar in that they are both solid forms of water created at freezing or below freezing temperatures. They are different in that snow forms on minute particles in a cloud and may fall to the ground, whereas frost forms directly on objects on the ground and does not fall through the sky. When snow forms, water vapor changes into a solid or frozen form without first becoming a liquid. This process is known as sublimation. Frost forms when water vapor freezes directly on an object. When water vapor condenses, or comes out of the air on an object as a liquid, the liquid formed is called dew.

Experience: Part I

PURPOSE: To discover that there is water vapor in the air, even though it cannot be seen.

To discover that cold things can bring water vapor out of the air.

MATERIALS: One glass of warm water; one glass of ice water; food coloring; a white paper towel.

The glass of ice water should be allowed to sit at room temperature until water has condensed on the sides of the glass. Let the children feel the glass in order to confirm that there is water on the outside; then ask if they know where the water came from. (They usually suggest that it came from the inside of the glass.) Let them wipe all the water off the outside of the glass, and then mix a few drops of food coloring into the water. While condensation is forming again on the sides of the glass, ask the children what color they think the water should be on the outside of the glass now if, as they suggested, the water comes from the inside of the glass. Let them wipe the water off the glass with the white paper towel in order to see what color it is. Discuss what there is in the room (around the outside of the glass) that is the only other thing from which the water could have come. Let the children feel the glass of warm water in order to confirm that there is no water on the outside of this glass. Discuss what kinds of things are able to bring water out of the air (warm things or cold things).

Suggested Questions

1. What is on the outside of this cold glass?
2. Where do you think the water on the outside of the glass came from?
3. What color do you think the water on the outside of the glass will be now that we've colored the water in the glass?
4. After the children have wiped the water off the glass, ask, "What color was the water? Did it come from the inside of the glass?"

5. What is there all over this room that is around this glass?
6. Do you think the water must have come from the air?
7. What kind of things can bring water out of the air (cold or warm)?

Experience: Part II

PURPOSE: To learn that when water vapor comes out of the air on something so cold that the water vapor turns to a solid form, it is called frost.

MATERIALS: An empty glass; an electric freezer.

Let the children feel the glass in order to confirm that it is neither wet nor extremely cold. Place the glass in the freezer and leave it for at least thirty minutes. Remove the glass, and within a minute, frost will form on the sides. Tell the children that what has formed on the glass is called frost. Let them scrape some of the frost off the glass, and then confirm that the glass became very cold in the freezer. You may wish to compare the formation of frost to the formation of dew (the liquid that condensed on the glass in Part I was dew). It may be necessary to allow the frost to melt into a liquid form of water in order for the children to understand that frost is also water that has come out of the air. When children can "see" their breaths on a very cold day, they are seeing the result of water vapor from their warm, moist breaths condensing on cold particles in the air.

Suggested Questions

1. After removing the glass from the freezer, ask, "What is on this glass? How does this glass feel?"
2. After allowing the frost to melt, ask, "What is frost made of?"

3. Where did the frost (or the water that made the frost) come from?
4. What brings frost out of the air?

Experience: Part III

PURPOSE: To discover that snow is made of water.
MATERIALS: A pan to hold snow.

Let the children fill a pan with snow and bring it indoors to watch it melt into water.

Suggested Questions

1. What is snow made of?
2. Was it cold outside where the snow was?
3. What made the snow melt or turn to water?

Experience: Part IV

PURPOSE: To observe what water vapor is and to learn, in part, how water evaporates.

To learn that snow crystals may begin as water vapor.

MATERIALS: A pan containing about one cup of water; a source of heat.

Boil the water while the children observe it disappearing into the air. Tell them that this water in the air is called water vapor. Explain that snow can be made of water vapor—water that has disappeared into the air.

Suggested Questions

1. What is happening to the water? Where did it all go?
2. Must the temperature be warm or cold for the water vapor to be snow? (What happened to the snow in the previous experience when it was warmed?)

Experience: Part V

PURPOSE: To discover that snow forms around tiny particles (perhaps dust or soot) in the air.

MATERIALS: The pan of melted snow from Part III.

Let the children look at the pan of melted snow in order to observe the dust or soot particles floating in it. Recall that in Part I and Part II water came out of the air onto the glass. See if the children are able to suggest that the water that became the snow could have come out of the air onto some of these particles. Note: Some of the freezing nuclei for snow are microscopic, and some of the particles seen in the water were collected as the snow fell.

Talk about some things other than dust or a cold glass—for example, leaves, grass, windows, and roofs—that typically bring water out of the air.

Suggested Questions

1. What do you see floating in the water?
2. What do you think brought the water out of the air to make snow? Onto what did the water that made the snow come out of the air?
3. Where does snow come from? Does it fall from the sky, or does it just appear on things, like frost or dew?

4. Where must the particles be that bring water out of the air to make snow?
5. Can you think of some other things that bring water out of the air?
6. What are some things that you have seen frost or dew on?

Follow-Up Activities

Have the children scrape some frost from a window pane and let it melt. Discuss what the frost was made of and what brought the water out of the air.

Let the children collect some snow by allowing it to fall on a piece of dark-colored cloth, preferably velvet or felt that has been wrapped around cardboard. Black construction paper may be used, but it does not work as well as a soft cloth, such as velvet or felt. Let the cloth-covered cardboard or black construction paper cool outside or in the freezer before catching the snow. If they look through a very strong magnifying glass or a microscope, the children will be able to see the six sides of a snowflake crystal.

Let the children make a six-sided paper snowflake. Fold a square piece of paper—first into quarters, and then into a cone. Cut off the top of the cone in a zig-zag fashion as shown in Figure 2-1; then cut jagged designs into the paper. Do not cut through

Figure 2-1
Paper Snowflake

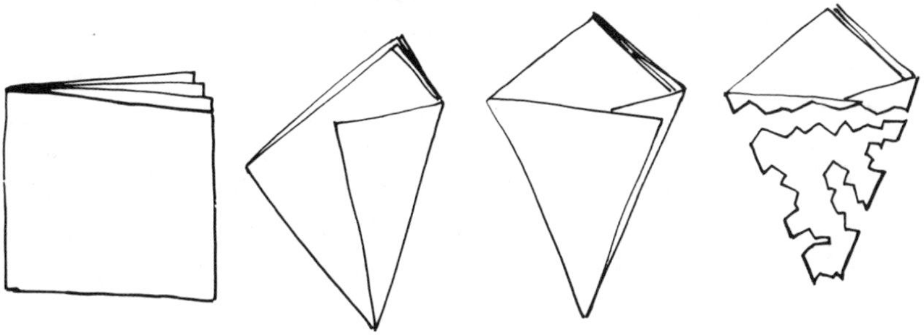

the full width of the cone. Open the paper in order to see a six-sided snowflake.

Let the children make snow cones by placing fresh, clean snow in a paper cup and pouring maple syrup, a mixture of vanilla flavoring and sugar, or concentrated juice over the snow.

Suggested Readings

Hurd, Edith Thacher. *It's Snowing.* New York: Sterling Pub. Co., 1957.

Keats, Ezra Jack. *The Snowy Day.* New York: Viking, 1962.

CURRENT ELECTRICITY

Information for Adults

Electricity is, in a sense, a force much like the wind: It cannot be seen, although what it does may be seen. Scientists do not know exactly what electricity is, although they do know a great deal about how it works. Electricity is made by infinitesimal particles called atoms. These atoms, which are so small that they cannot be seen, are made up of a core that consists of protons and neutrons and an outer shell that consists of orbiting electrons. The electrons are essentially responsible for creating electricity.

Because electrons are moving along the outer edge of the atom, they may easily move to another atom or be exchanged with another atom. Most things keep a balance or equal number of protons and electrons, so that if an extra electron were added to an atom, the atom would get rid of another electron in order to retain a balance. If atoms were lined up in a wire and an extra electron were added to the first atom, this atom would pass an electron to the second atom in order to regain balance; the second atom would then have to pass on an electron to the third atom, and so on down the wire. It is this passing or moving of electrons that creates current electricity.

The extra electron that starts the movement comes from a battery or a generator. Actually, many electrons are sent out at one time, so that there are many atoms involved in exchanging electrons. A generator is usually able to send out many more electrons than a low voltage battery; hence it is able to produce a more powerful electrical force. A battery starts the movement of electrons with chemicals, whereas a generator uses coils of wire and magnets (when a loop of wire is passed through the "lines of force" of a magnet, electrons begin racing through the wire).

Electrons are able to travel only in conductors—certain materials, such as metals (particularly copper and aluminum). Materials that will not allow electrons to travel are called insulators and include wood, porcelain, glass, plastic, rubber, and cotton. Insulators are used in order to prevent the unwanted flow of electricity.

Electrons producing a direct current require a complete circuit or circle in which to travel, always returning to the place where they started. An alternating current, which is the type of electricity in most homes, may or may not require a complete circuit. In both types of current, the electricity flow may be stopped by a switch, a place where the circuit is broken.

The study of electricity may be introduced through the use of a circuit or electricity board. For this, the adult will need one board measuring approximately one foot by one foot, one miniature porcelain light receptacle with an appropriate size of bulb, two wood screws, two finishing nails, about three feet of uncoated or uninsulated copper wire, and one six-volt battery (dry cell). All of these materials may be purchased from a hardware store or in a kit from a teacher's supply store. See Figure 2-2 in order to determine how materials should be organized to create an electricity board.

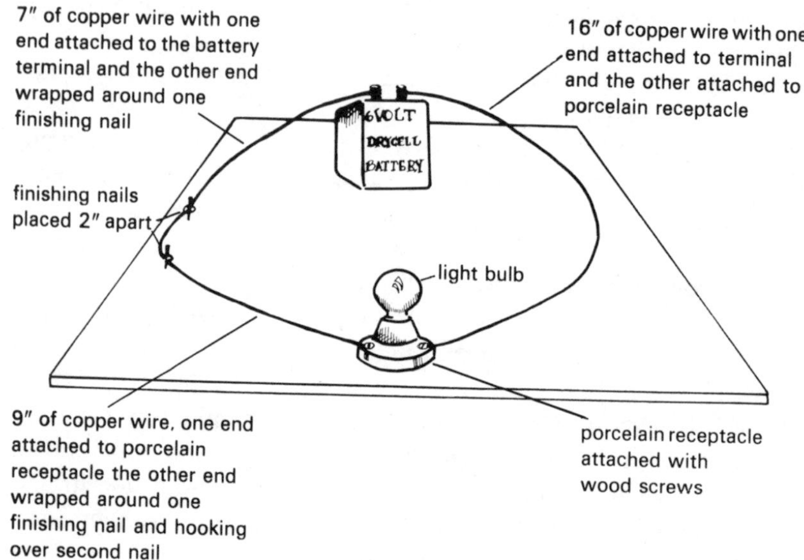

Figure 2-2
Electricity Board

Experience: Part I

PURPOSE: To learn that we can get electricity from a low voltage dry cell battery.

To observe that electricity travels through wires in a circular pathway called a circuit.

To discover that electricity gives us heat and light.

MATERIALS: The electricity board described in "Information for Adults."

With the battery disconnected from the electricity board, show it to the children and tell them what it is, if they don't already know. Explain that we get electricity that does not hurt us from a battery like this.

Place the battery and the electricity board out so that the children may experiment to discover how the battery might make the light come on. Ask what caused the light to come on and what we get from a battery.

Suggest that the children find what the electricity traveled through to get from the battery to the light. Ask them to use a finger to follow the pathway of wire that the electricity traveled in. Explain that this pathway is called a circuit.

Let the children feel the light bulb in order to confirm that electricity gives us heat or warmth as well as light.

The adult may wish to make it clear that the electricity that is in our homes can hurt us if we don't use it properly, and that children must not experiment with it.

Suggested Questions

1. What made the light come on? What did the electricity do to the light?
2. In this experience, through what did the electricity travel?
3. In this experience, what shape was the circuit or pathway that the electricity followed?
4. What does electricity give us other than light?

Experience: Part II

PURPOSE: To discover that when the circuit is broken or incomplete, the electricity will not work.

To learn that the place where we can break or complete a circuit is called a switch.

To learn that electricity can travel through some materials but not through others.

MATERIALS: The electricity board; small pieces of wood, plastic, string, rubber, glass, and aluminum; a nail. (Each of these items should be long enough to reach the full length between the two nails on the electricity board.)

Leave the wires attached to the battery, but let the children "unhook" the piece of wire between the nails in order to see what happens. Using a finger, have them trace the pathway of the circuit in order to discover that the circuit is now broken or incomplete. Explain that the piece of wire that can break or complete the circuit is called a switch. See if they are able to find another switch in the room and break and complete a circuit with that switch.

Let the children experiment with the items listed in "Materials," placing them between the nails (touching both nails at the same time) in order to discover which of the items will conduct electricity. The children may classify these items into two groups, conductors and insulators. The adult may wish to discuss why the wires in our houses and those connected to our electric appliances are insulated with plastic or rubber.

Suggested Questions

1. What happened when you removed the wire that was between the nails?
2. What did this do to the circuit or pathway?
3. Can you find a switch in the room and use it to break and complete the circuit? What happens to the lights when you break and complete the circuit?
4. What things will electricity travel through? What will it not travel through?

Experience: Part III

PURPOSE: To learn, in part, how electricity is made.

MATERIALS: One penny for each child; two pennies for the adult.

Have the children sit in a circle, which represents a circuit. Each child represents an atom (a very small part that makes up the circuit). Each "atom" has a penny, which represents an electron. The adult, who should be sitting with the children to help form the circle, represents the starter (a battery or generator) and gives his or her extra "electron" (extra penny) to one child. The adult explains that no "atom" may give an "electron" back to the "atom" who gave it to him or her and that an "atom" may not keep more than one penny (electron) at any time. The children will discover that they must pass the extra electron on to the next child so that the electron moves around the circle, back to the starter, and begins around the circle again.

Suggested Questions

1. We are sitting in a shape that represents a circuit. What shape is this circuit?
2. Who or what caused the electrons to begin moving around the circle or circuit?
3. What happened when the extra electron came back to me (the starter)?
4. In a real circuit, what causes the electrons to begin moving through the wires? (Where did the electricity come from on our electricity board?)

Follow-Up Activities

Discuss with the children some of the many ways that we use electricity.

Cut away the insulation from the cord of an old appliance so that the children may see that it contains not one but two wires. Shape the two wires into a circle so that the children may see that they form a circuit. See if the children can explain why there were two wires and why the wires were covered or wrapped.

A small bell or buzzer purchased from a hardware store

may be used to replace the light on the electricity board in order to illustrate still another way that electricity may be used.

As another example of the way that electricity may be used, help the children make electromagnets. For directions see "Experience: Part V" in "Magnets."

Suggested Readings

Bradbury, Ray. *Switch on the Night.* New York: Pantheon, 1955.

Branley, Franklin, and Vaughn, Eleanor. *Rusty Rings a Bell.* New York: Thomas Y. Crowell, 1957.

Scarry, Richard. *What Do People Do All Day?* New York: Random House, 1968 (see pages 14-15, 72-73).

Victor, Edward. *Electricity.* Chicago: Follett Publishing Co., 1967.

STATIC ELECTRICITY AND FRICTION

Information for Adults

Everything is made up of electrical particles called protons and electrons, both of which are so small that they cannot be seen. Scientists say that a proton carries a positive charge (+) and that an electron carries a negative charge (-). If not charged, most things carry an equal number of each; the positive and negative charges are balanced.

Sometimes, when two objects are rubbed together vigorously, friction can cause electrons (which are less stable than protons) to be rubbed off: One of the objects being rubbed loses some of its electrons, and the other object gains these electrons. The protons and electrons become unbalanced; the objects are said to be charged; and static electricity has developed. Some things are more likely than others to develop static electricity, as it is easier for them to gain or lose electrons.

When presenting this unit, choose a very dry clear day, because electrical charges will leak off onto droplets of water in the air on damp days, and static electricity will not easily develop. The adult may wish to wait for a day when static electricity is obviously building up on objects.

Experience: Part I

PURPOSE: To learn that when two things are rubbed together vigorously, they may create friction.

To discover that friction makes things feel warm.

MATERIALS: A smooth surface, such as a counter top or table top; two small pieces of smooth metal; a box of matches.

Tell the children to rub their hands together *very* hard and very fast. Ask how their hands feel now (warm or cold). Tell them that when two things are rubbed together very hard and very fast, they make friction.

Tell the children to rub one hand very hard and fast on the counter top or table top. Let them compare the warmth of the spot where they rubbed to the coolness of the rest of the counter.

Let the children feel the pieces of metal in order to establish that they feel cool; then rub them together vigorously (it may be necessary for the adult to do this in order to create sufficient warmth). Let the children feel the pieces of metal again in order to establish that friction has made them feel very warm.

Let the children feel the match box and the matches in order to establish that they are cool (or not hot). Strike several matches on the side of the box and let the children feel the spot. Ask how they think the matches feel. It may be necessary to discuss some safety rules regarding matches (children should not play with matches and should not use matches even in "science experiments" unless an adult is supervising).

Suggested Questions

1. How do your hands feel after you have rubbed them together (warm or cold)?
2. How did the counter feel after you rubbed your hand over it? What did you make?
3. How did the pieces of metal feel after we rubbed them together? What did we make?
4. How did the match box and matches feel when we rubbed them together? What did we make?
5. How do you make friction? How does friction make things feel?

Experience: Part II

PURPOSE: To learn that when something is rubbed vigorously, friction may cause it to become charged (static electricity may develop).

To observe that static electricity can make things stick together.

MATERIALS: Pieces of plastic wrap (approximately 3" × 3"); a hairbrush; some very small pieces of paper.

Have the children place a piece of plastic wrap under the palm of one of their hands, and then have them release the plastic wrap to see it fall to the floor. Now have each one rub the piece of plastic wrap against a sweater and again place it under the palm of one hand and release it in order to observe that this time it sticks to the hand. The adult may tell the children that they charged the pieces of plastic wrap and made static electricity when they rubbed the plastic wrap against their sweaters.

Give a child a hairbrush and ask him or her to see if the small pieces of paper will stick to it. Have the child brush his or her hair as, together, you count twenty strokes. Now let the child see if the paper will stick to the brush.

Suggested Questions

1. What are we making when we rub two things together very hard and fast?
2. What did you do in order to make the plastic stick to your hand?
3. What did you do to make the paper stick to the brush?
4. When something is rubbed to make it charged, what might it do to other things?
5. What do we call it when something is rubbed, becomes charged, and makes things stick to it?

Experience: Part III

PURPOSE: To observe that when two things are charged alike, they push each other apart.

To learn that similarly charged things repel or push each other apart.

To learn that things with opposite charges attract or stick to each other.

MATERIALS: Two inflated balloons for each child (these should have strings attached); two child's snap-together beads of the same color and shape.

Ask the children what they think would happen if they rubbed the two balloons against their sweaters and then let go of the balloons. Let them try this experiment; then ask what they have made or built up on the balloons by rubbing them.

Let the children rub both balloons at the same time against their sweaters. Then place the two balloons side by side on a table top or on the floor in order to observe that the balloons push each other apart. Charge the balloons again and have the children try to bring the balloons together while holding only the string of each balloon. This may be repeated using other similar objects, such as two pieces of plastic wrap or two wool sweaters. Hold these objects next to each other, one in each hand, in order to see them push each other apart.

In order to begin to help the children understand why two things with unlike charges attract each other, first establish that the two balloons (or the two pieces of plastic wrap) are like each other but are different from the sweater. Suggest that perhaps these differences make things charge or build up static electricity differently. Now compare these objects to the two snap-together beads, which are different on each end, even though the two ends of one are like the two ends of the other. Have the children stick the beads together in order to see which ends attract or stick to each other.

Suggested Questions

1. What happened to the two balloons (pieces of plastic wrap, etc.) after they were rubbed to make static electricity and then placed next to each other?
2. Are the two balloons alike? Are the two pieces of plastic wrap alike? Are these different from each other or from the sweater?
3. Were the two ends of the snap-together beads that stuck together alike or different?
4. Will the two ends of the beads that are alike stick together?
5. Do things have to be charged alike or differently in order to stick together?

Experience: Part IV

PURPOSE: To discover that static electricity can build up to create sparks, which are like little flashes of lightning.

To learn that these sparks jump from the charged object to the uncharged object.

MATERIALS: A wool or nylon carpet or remnant; a darkened room; a metal object, such as a doorknob or a pencil sharpener.

Ask the children whether they think static electricity could build up on their bodies. Have them rub their feet vigorously against the carpet or remnant, then have them touch the metal object in order to observe what happens. Ask what they made when they rubbed against the carpet, and then ask what this caused to happen.

Let a child rub his or her feet again on the rug, and then ask the child to touch you (you have not been rubbing your feet to build up static electricity). Ask whether the sparks jumped from the electrified or the unelectrified person.

The adult may wish to mention that these sparks may be compared to flashes of lightning. Sometimes static electricity builds up inside clouds; then large sparks jump from the clouds that are electrified to something that is not electrified, such as a tree or a house.

Suggested Questions

1. What happened when you rubbed your feet against the carpet and then touched the metal object?
2. What did you make in your body when you rubbed your feet against the carpet?
3. From where did the spark jump? To where did it go?
4. Were you electrified (did you have static electricity in your body)? Was the metal object electrified?

Follow-Up Activities

Let the children experiment to see what happens when two children scuff their feet across the carpet and then touch each other. Why was there no spark? Suggest that the children look for static electricity on a cool, dry day when their mothers remove clothes from the dryer, when they brush or comb their hair, when they are taking off sweaters or coats, and when they walk across the carpet and open the door, using the metal doorknob.

Suggested Readings

Schwartz, Julius. *Now I Know.* New York: McGraw-Hill, 1955.

Zolotow, Charlotte. *The Storm Book.* New York: Harper & Row, Pub., 1952.

LIGHT AND SHADOWS

Information for Adults

There must be a source of light for an object to be visible. Some objects create their own light (the sun, a candle, a light bulb), but most objects create no light and may be seen only because of the light that is reflected from them. A ray of light that travels in a straight line hits an object and is "bounced" or reflected off that object to another object or to the eye. Sometimes the rays of light bend or change direction if they hit an object at an angle other than a right angle. Certain mediums, such as water, can change the direction of a ray of light as it passes through the medium. Light bent in this manner is said to be refracted.

Objects with smooth, shiny, or light-colored surfaces are able to reflect a great deal of light. Objects with rough, dull, or dark-colored surfaces absorb most of the light and reflect very little. Something that appears white in color is reflecting most of the light, whereas something that appears black in color is absorbing most of the light.

Darkness is created by the absence of light. Materials that let most of the light shine through them are said to be transparent; materials that let only some of the light shine through them are said to be translucent; and materials that let no light shine through them are said to be opaque. When a ray of light falls on an opaque object, an area of darkness that is the same shape as the object that blocked the passage of light is created. This area of darkness (a shadow) will appear to change in size according to the angle at which the light rays fall on the opaque object. The shadow moves only if the object creating the shadow moves or if the source of light moves.

Light is energy created by atoms in the form of photons. Different kinds of atoms give off different kinds of photons, each of which has a different amount of energy. It is this energy that creates color. The color spectrum begins with the color created by low energy photons (red) and moves to the color created by the highest energy photons (violet); the order of the spectrum is red, orange, yellow, green, blue, violet.

Experience: Part I

PURPOSE: To observe that light travels in a straight line.

To discover that light may be "bounced" or reflected.

To learn that we are able to see an object when it reflects light.

MATERIALS: A projector light or bright flashlight; two chalkboard erasers filled with chalk dust; a ball; one or more mirrors; a very dark-colored object.

Darken the room and turn on the projector or flashlight. Let a child clap the erasers together in the beam of light. The dust will make the beam of light more distinct, and the children may follow the outline of the beam with their fingers in order to find that light travels in a straight line.

Let a child throw the ball against the wall and describe what happens. Let the child see what happens to the path of the ball when it is thrown toward the wall at different angles. Ask the children whether they think a beam of light can bounce like this ball. Let them use the mirror to experiment with reflecting light. Two mirrors may be used to "bounce" the same light beam in different directions, and the chalk dust may be used to help see the light beam more clearly.

In a very dark room, hold up a dark object and ask the children whether they are able to see the object. Ask if they are able to suggest what might be done in order to help them see the object better. Let them use the flashlight to "bounce" or reflect light off the object.

Suggested Questions

1. In what kind of line does light travel?
2. Can light bounce off an object like a ball bounces off a wall?
3. How can we change the direction in which light bounces?
4. What can you do with light to help you see an object better?

Experience: Part II

PURPOSE: To observe that some surfaces reflect light better than others.

MATERIALS: A mirror; two pieces of aluminum foil (one smooth; one crinkled); one piece each of black and white paper; a flashlight.

In a darkened room, the adult should hold the flashlight at a slight angle in order to reflect light from each of the surfaces listed in "Materials." Each child, one at a time, will place a hand about 36 inches above each different surface. The children will be able to see their hands more clearly when the surface reflects light well.

Suggested Questions

1. Is this surface reflecting light (can you see the light beam "bounce")?
2. Which surface reflected more light to help you see your hand better?

Experience: Part III

PURPOSE: To discover how a shadow is made.

To observe that a shadow is the same shape as the solid or opaque object that stops the light.

MATERIALS: A projector or bright flashlight; a movie screen, a sheet, or a light-colored wall; a piece of clear plastic and a piece of black paper (each about 8″ × 10″); some opaque objects (including cutouts of squares, circles, and triangles).

Shine the light on the screen, sheet, or wall. Let the children hold the plastic in front of the light near the light source in order to see that the light goes through it; then let them hold the black paper in front of the light near the light source in order to see that the light does not go through it. Ask what the paper made on the wall, and if they don't know what this "darkness" is called, tell them that it is a shadow. Let them use other opaque objects, including the shape cutouts, to make shadows and, in each case, ask what shape the shadow is.

Turn off the light and ask the children to try to make a shadow. Turn the light back on and let them hold an object behind the light source to try to make a shadow.

Suggested Questions

1. Did the paper let the light shine through it or did it stop the light?

2. What kinds of things can make a shadow (those that let light shine through or those that stop the light)?
3. What shape is a shadow, or what gives a shadow its shape?
4. Can the shadow of a square ever be the shape of a triangle, or can the shadow of a pencil ever be the shape of a book?
5. What happened to the shadow when we took the light away?
6. What happened to the shadow when the object did not block or stop the light?
7. What things are necessary in order to make a shadow?

Experience: Part IV

PURPOSE: To discover that a shadow can move only if the light source or the object blocking the light moves.

To discover that the size of a shadow changes as the angle and the distance of the light from the object change.

MATERIALS: A bright flashlight; a movie screen, a sheet, or a light-colored wall.

Let the children experiment with the flashlight and one hand to discover two ways to make a shadow move.

Suggest that each child hold the light closer to one hand and then farther away from that hand in order to see what happens. Show the children how to change the angle of the light to make the shadows of their hands grow larger or change in appearance.

Take the children outside in the early morning and help them measure the height or length of their shadows. Repeat this activity around noon. Have the children notice where the sun is each time.

Suggested Questions

1. What are two ways that a shadow can be made to move?
2. What happens as the light gets closer to your hand? As it gets farther away?

3. When was your shadow smaller or shorter? When was it taller or longer?
4. From where did the light come to make a shadow when you were outside?
5. Can you think of what might have caused your shadow to change in size when you measured it outside?

Follow-Up Activities

Allow the children to experiment with their shadows outside. Can they catch their shadow, step on it, or run away from it?

Play a "Guess What I'm Doing" game with shadows. Set up a lamp or projector light behind a sheet that is hanging in the room or across a doorway. Let one child stand between the sheet and the light and perform some activity. The children who are sitting on the opposite side of the sheet must guess what this child is doing by watching the shadow.

A puppet show may be presented with shadows. Set up a light and a sheet as suggested in the above activity and use simple silhouettes cut from cardboard attached to drinking straws (which act as handles). Use a cardboard box as the stage for the children to present a familiar story or fairy tale.

For an art project, the children may make shadow pictures. Have each child draw a simple picture (one or two objects—a house, a tree, themselves, etc.) with crayons on a half-sheet of white construction paper. Place the picture on top of a half-sheet of black construction paper and glue or staple the two papers together at the four corners. Now the children should cut out the objects that they have drawn (at the same time they will be cutting a duplicate outline of each object from the black paper). On a full-sized piece of white construction paper, the children may glue each object and its black "shadow," placing the bottom edge of the object and the shadow together so that the shadow forms a mirror image of the object. Suggest that in this picture they now draw the sun, which would be the source of light for making the shadows in the picture.

Suggested Readings

Alexenberg, Melvin. *Light and Sight.* Englewood Cliffs, N.J.: Prentice-Hall, 1969.

Bradbury, Ray. *Switch on the Night.* New York: Pantheon, 1955.

Branley, Franklyn M. *Light and Darkness.* New York: Thomas Y. Crowell, 1975.

Johnson, Crockett. *Will Spring Be Early?* New York: Thomas Y. Crowell, 1959.

Lawery, Lawrence. *Dark is a Shadow.* New York: Holt, Rinehart, & Winston, 1969.

Leaf, Munro. *Boo.* New York: Random House, 1948.

Pondendorf, Illa. *Shadows and More Shadows.* Chicago: Children's Press, 1971.

Simon, Seymour. *Let's Try-It-Out: Light and Dark.* New York: McGraw-Hill, 1970.

ROCKS

Information for Adults

There are three categories of rocks: (1) Igneous rock, made by fire and heat, was once melted minerals inside the earth. This liquid worked its way to (or near) the surface, where it cooled, becoming solid to form rock. Pumice, granite, and basalt are igneous rocks. (2) Sedimentary rock is layered rock formed by pressure. It is usually formed when dust, dirt, sand, gravel, mud, and pieces of other rocks or seashells are washed to the bottom of oceans, lakes, ponds, or rivers and become pressured and cemented together by the water. Shale, sandstone, and limestone are sedimentary rocks. (3) Metamorphic rock is changed rock. It was once sedimentary or igneous rock and has been changed to metamorphic rock by pressure or heat. Slate (which began as shale), quartzite (which began as sandstone), marble (which began as limestone), and gneiss (which began as granite) are metamorphic rocks.

All rocks are made up of two or more of the approximately 2,000 different known minerals. Rocks look different because of the way they were formed and because of their mineral content. With the children, the formation of rocks might be compared to a recipe used for cooking. When making vanilla refrigerator cookies and chocolate chip cookies, the recipes call for different ingredients and for different methods of preparation. Ingredients and the method by which a rock is made determine differences in appearance. Rocks will have different colors, textures, and patterns, according to their recipes.

The exteriors of rocks also differ in appearance, according to the conditions to which they have been exposed. Rocks that are relatively new pieces broken from larger rocks or mountains have rough outlines and sharp edges. Rocks that have been in rivers or streams will be smoother and have rounded edges because of the wearing effect that running water has on the rocks.

Rocks may be classified according to their mineral content. A "scratch test" may be used in order to determine hardness of minerals. Geologists use the MOH Scale to classify minerals according to hardness. This scale, named for Friedrich Mohs who devised it, consists of a list of ten items, each ranked by number on a scale from one to ten. If one of the items is able to scratch a mineral in a rock, that mineral is given a number from the scale that corresponds to the item that was able to make the scratch. The hardest mineral would be given a score of ten, and the softest mineral would be given a score of one. For the children, rocks may be grouped according to hardness by using a fingernail to determine "soft" rocks, a ten-penny nail to determine rocks "medium" in hardness, and a steel nail file to determine "hard" rocks. A rock is rated according to which is the first of these items to make a scratch (not just a mark) on it.

The "streak test" may also be used to classify minerals in rocks as metallic or nonmetallic. The unglazed side of a piece of ceramic bathroom tile may be used in this test. Mark on the tile with a rock: If it leaves a black, dark brown, grey, or dark green streak, the rock contains metal; if the streak is white or clear, the minerals in the rock are nonmetallic.

Minerals in rocks that are classified as carbonates may be determined by an "acid test." Place a rock in vinegar (a mild acid): If the rock fizzes (as bicarbonate of soda fizzes in vinegar), it does contain carbonates; if the rock does not fizz, it does not contain carbonates. Some rocks may have to be crushed or cracked open in order for the fizzing to be obvious.

Some commonly found rocks and the minerals that they contain are as follows:

Limestone (used in building)—Calcium carbonate is the main mineral, although others may be present.

Marble—Calcium carbonate plus graphite (if streaked with black). Calcium carbonate plus chlorite (if streaked with green). Calcium carbonate plus iron (if streaked with pink).

Granite—Feldspar, quartz, and mica (if silver or shiny specks are present) as well as hornblende (if black, black-green, or brown specks are present).

Much of the soil was once rock that has been broken into very small pieces and worn down by wind, rain, the heat of the sun, frost, and running water. These weathering factors convert larger pieces of rocks into grains of sand or soil.

Experience: Part I

PURPOSE: To learn that rocks may be classified or grouped according to the hardness of the minerals they contain.

MATERIALS: A collection of rocks; small boxes labeled hard, medium, and soft; a ten-penny nail; a steel nail file. If desired, picture cards showing a fingernail, a ten-penny nail, and a steel nail file may be placed on the boxes (the steel nail file on the box labeled "hard," the ten-penny nail on the box labeled "medium," and the fingernail on the box labeled "soft").

Bring in a collection of rocks that you have gathered or let the children gather rocks. Tell them to place the rocks into groups; then ask how they decided to put the rocks together in that way (color, shape, size, etc.).

Show the children how to do the "scratch test"; then ask if they can suggest another way to classify or group their rocks. Allow them to reclassify the rocks according to hardness.

Suggested Questions

1. Why did you group your rocks in this way?
2. Can you tell me some other way in which you might have grouped the rocks?
3. What object was able to scratch this rock (a fingernail, nail file, ten-penny nail)?
4. Was the rock soft, medium, or hard? In which box does it belong?

Experience: Part II

PURPOSE: To learn that rocks may be classified or grouped according to whether they contain metallic or nonmetallic minerals.

MATERIALS: A collection of rocks (be sure that some of these contain metal); pieces of ceramic tile with one side unglazed; a chart or pieces of tile showing the colors that indicate metals and nonmetals in the "streak test." (These colors are listed in "Information for Adults.")

Demonstrate the "streak test" and show how to match the colors obtained with the "streak test" to the colors on the chart or on the tiles marked with metallic and nonmetallic minerals. Now ask the children if they are able to suggest yet another way in which rocks may be grouped. Allow them to classify their rocks according to metal content.

Suggested Questions

1. This rock contains a metal. What color streak does it make on the tile?
2. This rock does not contain a metal. What color streak does it make on the tile?
3. Does this rock contain metal? How about this one?
4. Can you think of another way that you might group your rocks?

Experience: Part III

PURPOSE: To learn that rocks may be classified according to whether they do or do not contain minerals called carbonates.

MATERIALS: A collection of rocks, some of which contain carbonate minerals (rocks containing carbonates may be found near building sites or near the seashore, or you may use common chalk); a bowl or glass of vinegar.

Demonstrate the "acid test" by dropping a rock that contains carbonates then one that does not contain carbonates into the vinegar. Explain that only the rocks that fizz contain minerals called carbonates. Ask the children if they are able to suggest still another way in which rocks may be grouped. Let them classify their rocks according to carbonate content.

Suggested Questions

1. This rock contains carbonates. What does it do when I drop it into the vinegar?

2. This rock does not contain carbonates. What does it do when I drop it into the vinegar?
3. Does this rock contain carbonates? How about this one?

Experience: Part IV

PURPOSE: To understand that all rocks are made up of two or more minerals. To examine some of the minerals that make up rocks.

MATERIALS: A collection of rocks; a commercial mineral specimen kit (obtainable from a hobby shop, lapidary shop, or teacher's supply store).

Ask the children why they think rocks vary in appearance. Show the mineral specimen kit, explaining to the children that their rocks contain some combination of these minerals. Help them to find the combinations of minerals that make up their rocks by trying to match the appearance of the minerals they see in their rocks with the appearance of the minerals in the specimen kit. Discuss differences in appearances of rocks in terms of the recipe analogy (see "Information for Adults").

Suggested Questions

1. Do two rocks that contain the same minerals look alike or different?
2. Do two rocks that contain a different combination (or group) of minerals look alike or different?
3. Why do you think rocks look different?

Experience: Part V

PURPOSE: To discover what rocks look like on the inside. To learn that weathering changes the appearance of a rock.

To discover that the wearing away of rocks helps make the soil.

MATERIALS: Some rocks (some rough; some smooth); a towel; a hammer; a piece of white paper.

Wrap a smooth rock in the towel and let the children crack it open with the hammer in order to observe new colors, patterns, and textures—even crystals—on the inside. Point out that the new smaller rocks that came from cracking the larger smooth rock are rough in texture. Discuss how weather can change rocks.

To discover what else weathering can do, let the children rub two rocks together while holding them over the piece of white paper. They should be able to observe small pieces of soil falling on the paper.

Suggested Questions

1. What did you find inside the rock when you broke it open?
2. Were the new little rocks that you made smooth or rough?
3. What, in nature, could cause a rough rock to become smooth?
4. What do you see falling on the white paper when you rub the rocks together?
5. What, in nature, would "rub" the rocks to make soil?

Follow-Up Activities

The unit on crystals (see pages 89–93) would be an excellent follow-up to explain how crystals are formed in rocks.

Discuss some ways that people use rocks: chalk and graphite (pencils) for writing; slate for blackboards, floors, and roofs; marble for building and for statues; gypsum for plaster. Gold, coal, and iron also come from rocks.

Let the children use small smooth rocks with rounded edges to create rock people and animals. These may be glued together, allowed to dry, and then painted. For this activity, a supply of riverbed rock may be obtained inexpensively from a nursery or garden supply store.

Suggested Readings

Baylor, Byrd. *Everybody Needs a Rock.* New York: Scribner's, 1974.

Podendorf, Illa. *The True Book of Pebbles and Shells.* Chicago: Children's Press, 1960.

——. *The True Book of Rocks and Minerals.* Chicago: Children's Press, 1972.

Ross, George. *What Did the Rock Say?* New York: Holiday House, Inc., 1970.

Wyler, Rose, and Ames, Gerald. *Secrets in Stones.* New York: Four Winds Press, 1970.

CRYSTALS

Information for Adults

Many of the rocks that children find will contain shiny particles or transparent gemlike structures (both are crystals, perhaps of quartz or mica). The children may ask what these "shiny" or "diamond things" are and how they got in the rocks.

Crystals are substances that take a definite shape, each type of crystal taking its own particular shape. Crystals have flat surfaces or faces that meet to form angles. The sharp edges created by the angles may be seen under a strong magnifying glass and help to identify a particular crystal. Some common crystals, many of which children will be familiar with, are table salt; rock salt; granulated sugar; rock sugar; sand; snowflakes; and gems, such as a diamond, topaz, ruby, or emerald.

Crystals may be formed in several ways: (1) when a substance is dissolved in a liquid, and the liquid then evaporates; (2) when a substance is dissolved in a hot liquid, and the mixture cools and solidifies; (3) when a substance is melted to a liquid state and then cools to a solid crystalline state. The appearance of a crystal will vary according to the substance or compound that has formed the crystal (not all substances or compounds form crystals). The formation of a crystal may be compared to a recipe: Each type of crystal has its own particular list of ingredients and its own method of preparation.

Experience: Part I

PURPOSE: To examine some common crystals and their formation.

MATERIALS: A strong magnifying glass; a piece of black paper; some common crystals (table salt, rock salt, rock sugar, sugar, sand, a rock containing crystal formations). Crystals may be observed in some of the rocks that the children crack open, or a rock containing crystal formations may be purchased inexpensively from a hobby shop, a rock shop, or a lapidary shop.

With the children, recall the shiny pieces or gemlike structures that they found inside their rocks (see "Rocks, Experience: Part V"). Explain that these are called crystals. Place some of the common crystals listed in "Materials" on black paper and examine them with a strong magnifying glass. Help the children to find the faces or flat surfaces of the crystals (count the faces) and to find the sharp angles. Note: If sand shows rounded edges, it is because the sharp edges have been worn smooth by water.

Suggested Questions

1. Do you know what we call the shiny pieces or the clear diamond-like pieces that you found in the rocks?
2. Can you find some flat sides or faces on this crystal when you look at it under the magnifying glass? How many faces does the crystal have?

3. When the faces come together, do they form a sharp, or a rounded edge?
4. Do all the crystals have faces or flat sides? Do they all have the same number? Do all the crystals have sharp or pointed edges?

Experience: Part II

PURPOSE: To make some crystals by dissolving a substance in a liquid and then allowing the liquid to evaporate.

MATERIALS: An aluminum pan or glass dish; pieces of coal or charcoal that have been cracked open; a spoon and bowl for mixing ingredients; 4 tablespoons of table salt; 4 tablespoons of water; 2 tablespoons of clear ammonia; 4 tablespoons of laundry bluing; plastic bottles of food coloring (optional).

Place the cracked coal or charcoal in the pan or dish, rough side up. Mix the salt, water, ammonia, and bluing together in the bowl, stirring well in order to dissolve as much of the salt as possible. Note: For safety reasons, the children should only crack the charcoal and help measure the salt and water in this activity. Pour the mixture over the charcoal and add drops of food coloring in spots if colorful crystals are desired. Place the dish where it will not be bumped and be sure to caution the children not to taste the mixture, as it may be quite harmful. In a few hours, crystals will begin to grow and will continue to grow for several days as the liquid evaporates from the pan. These crystals will more closely resemble sea coral than gems.

Suggested Questions

1. What do we call it when we make the salt disappear in the liquid?
2. What do you see growing on the charcoal? (Tell them that these are crystals if they don't know this.)

3. After a day or two ask, "What is happening to the liquid in the pan?"
4. When the liquid goes away, or evaporates, what is left behind?

Experience: Part III

PURPOSE: To make crystals by dissolving a substance in a hot liquid and then letting the mixture, or compound, cool and solidify.

MATERIALS: Three ounces of alum (this may be purchased inexpensively from a drugstore or grocery store); two cups of water; a pan; a source of heat; a glass or jar (this recipe may be cut in half, but fewer crystals will form).

Let the children measure the ingredients (alum and water), and then boil the mixture for approximately three minutes and pour it into a jar or glass. As the mixture cools, solid crystals will form on the bottom of the jar or glass. Do not disturb the mixture as the crystals form.

Suggested Questions

1. What happened to the alum when we boiled it in the water?
2. Is this mixture that we're pouring into the jar (or glass) hot or cool?
3. What do you see forming in the bottom of the jar (or glass) as the mixture cools?

Follow-Up Activities

Crystals may be formed by melting a substance to a liquid state and then allowing it to cool to a solid crystalline state. Purchase some salol (phenyl salicylate) from a drugstore. Place about two ounces of salol in a clear bottle with flat sides and melt it by

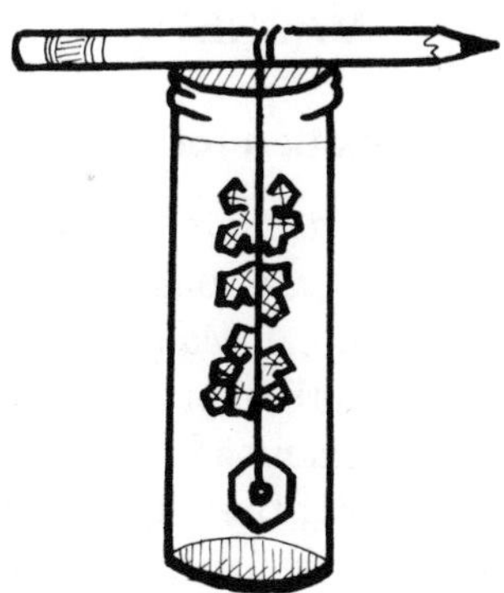

Figure 2-3

setting the bottle into a pan of boiling water. When the salol has melted, remove it from the heat and watch crystals form as the liquid cools to a solid state.

Let the children make crystals that they can eat (rock sugar candy). Boil one cup of water and two cups of sugar vigorously for one minute—to a temperature of at least 242° Fahrenheit on a candy thermometer or to the "soft ball" stage. Pour the mixture into a tall thin jar (such as an olive jar). Tie one end of a string that is approximately the length of the jar to a pencil and the other end to a small rustproof object that may act as a weight (a paper clip or a stainless steel nut may be used). Suspend the string in the center of the hot sugar water mixture and watch crystals form on the string as the mixture cools. In one or two days there will be enough crystals to eat. Do not disturb the jar while the crystals are forming (see Figure 2-3).

Suggested Readings

Arem, Joel. *Rocks and Minerals.* New York: Grosset & Dunlap, 1973.

Carona, Philip. *Crystals.* Chicago: Follett Pub. Co., 1971.

Zim, Herbert Spencer and Shaffer, Paul R. *Rocks and Minerals.* New York: Western Pub. Co., Inc., 1957.

MAGNETS

Information for Adults

There are two kinds of magnets, natural and manmade. Natural magnets, called lodestones, are rocks that contain large amounts of iron, a highly magnetic metal. Manmade magnets are made of a combination of highly magnetic metals, such as iron, steel, nickel, cobalt, or chromium. These manmade magnets may be made much stronger than natural magnets; they may be made into almost any shape but are usually made in bar shapes, horseshoe shapes, and cubes. A second kind of manmade magnet is an electromagnet, made by using electricity to magnetize iron.

A magnet has the unusual quality of attracting other pieces of iron, nickel, cobalt, or steel. The arrangement of atoms in a magnet makes it different from other iron. Every atom is surrounded by an electrical field that has a small bit of pulling power. In an ordinary piece of iron, the atoms—and thus the electrical fields—are arranged in a haphazard fashion: Each atom still has some pulling power, but, arranged in this manner, the atoms work against each other; one atom cancels out the other's pulling power. In a permanent magnet, all of the atoms are pointing in the same direction, working together to have great pulling power.

If an ordinary piece of iron or steel is stroked many times with a permanent magnet, always stroking in the same direction, the atoms in the iron or steel will be temporarily turned in the same direction, their pulling power working together in order to create a magnet. Eventually, the atoms will return to their original positions, and the iron or steel will lose its magnetic quality. In a permanent magnet, the atoms are permanently pointed in the same direction as the magnet is formed. However, if a permanent magnet is abused or exposed to too much heat, it may lose some of its magnetic quality.

Because of the arrangement of atoms in a magnet (all pointed in the same direction), all magnets have a north and a south pole. If allowed to swing or float freely, the north pole of a bar magnet will always point north, and the south pole will always point south. If a bar magnet were cut into many pieces, each piece would still have a north and a south pole, since cutting the magnet does not alter the arrangement of the atoms.

Around every magnet is a space in which the pull or force will work. This space, called a magnetic field, is stronger at the poles than at the center, and the pulling power of the magnetic field grows weaker farther away from the magnet. Because the earth has such a high iron content, it has a magnetic quality. In a sense, the earth is like a giant magnet with north and south poles, and it has a magnetic field that grows weaker as one moves away from the earth.

Experience: Part I

PURPOSE: To explore the pulling power of a magnet.

To observe that a magnet may pass on its magnetic quality to some other objects.

MATERIALS: A strong magnet; a variety of small objects, some of which a magnet will attract; pieces of paper, cloth, wood, glass, and aluminum foil; a bowl of water.

Place a variety of small objects on a table and let the children use the magnet to discover what it will attract (or what will "stick" to it) and what it will not attract. Let the children categorize the objects according to what the magnet will or will not attract.

Let the children further explore the pulling power of the magnet in order to see if it will attract a metal object through paper, cloth, wood, glass, aluminum foil, and water.

If the children have not yet discovered that magnetic power can pass through an object being pulled by a magnet and turn that object into a magnet itself, then demonstrate this to them and then let them experiment with this knowledge.

Suggested Questions

1. What does the magnet attract (or what sticks to the magnet)?
2. What does the magnet not attract?
3. Can the magnet attract an object through paper, cloth, wood, and so on?

Experience: Part II

PURPOSE: To discover that a magnet has a field of attraction surrounding it (called a magnetic field).

MATERIALS: A bar magnet; some small objects that a magnet will attract; a piece of white paper; some metal filings or bits of steel wool cut into very small pieces.

Suggest that the children place a magnet on the table and slowly move it near a metal object. Ask whether they can see the magnet start to pull the object toward it even before the magnet touches the object.

In order to let the children see the pattern or result of a magnetic field, sprinkle metal filings or bits of steel wool as evenly as possible over the surface of the white paper; then place the bar magnet on the center of the paper. The filings or steel wool will form an elongated circle around the magnet, with the accumulation of filings or steel wool being the greatest at the poles.

Suggested Questions

1. Does the magnet seem to have some power to pull an object even without touching that object?
2. What did the little pieces of metal or steel wool do when we placed the magnet near them?
3. Where did the most metal or steel gather?

Experience: Part III

PURPOSE: To discover that a magnet has a north and a south pole.

To explore the attracting and repelling ability of the poles.

MATERIALS: Two bar magnets with the north and south poles marked; a piece of string; a piece of paper with north marked on one end and south marked on the opposite end.

Let the children place the two magnets on a table and move the north or south pole of one magnet near the same pole of the other magnet in order to see that the magnets repel each other. Now move the north pole of one magnet near the south pole of the other in order to see the magnets attract each other.

Tie a string around the middle of one bar magnet (see Figure 2-4). Arrange the paper marked north and south, placing each end in the correct direction. (This may be done before the children are involved by using a compass or by allowing the magnet to swing freely from the string to determine north and south.) Now allow the children to suspend the magnet from the string, holding it over the paper marked north and south, as you explain what the markings mean. When the magnet stabilizes, note that the north pole is pointing to the north and that the south pole is pointing to the south (see Figure 2-4).

Figure 2-4

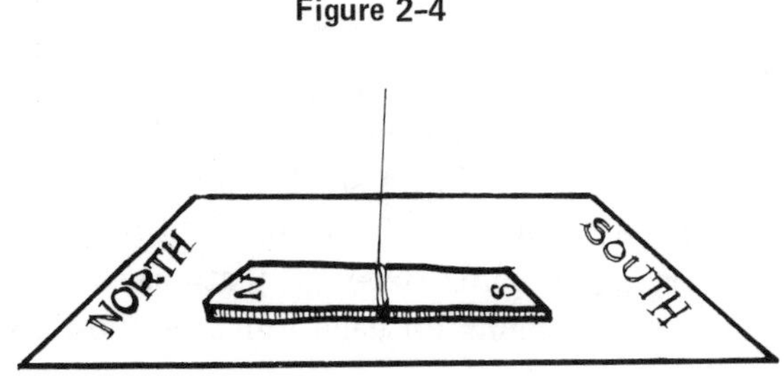

Let the children use the other bar magnet and the suspended magnet to experiment with the attracting and repelling ability of the poles.

Suggested Questions

1. What did the magnets do when you placed the two north poles or the two south poles together?
2. What did the magnets do when you placed opposite ends (north and south) together?
3. When we let the magnet swing freely, which direction did the north pole point to and which direction did the south pole point to?
4. Why do you think the ends of a magnet are named the north and the south poles?

Experience: Part IV

PURPOSE: To understand why a magnet has the ability to attract or pull.

To make a temporary magnet from a steel needle or an iron nail.

MATERIALS: A clear plastic medicine bottle containing metal filings; several small tacks or bits of steel wool cut into very small pieces; a magnet (or a "draw-a-picture" toy that contains metal filings and a magnet); a steel needle or an iron nail.

Explain to the children that magnets contain some things so very small that we cannot even see them and that these things are called atoms, each of which has a small amount of pulling power. Let two children be "atoms" and a third child be the object that the atoms are trying to pull. Have the "atoms" try to pull the other child by gently tugging on each arm in opposite directions.

(The child should be able to remain standing fairly stable.) Now have the "atoms" pull together gently on just one arm of the other child. (This time, the "atoms" should be able to pull the child toward them.)

In order to demonstrate how atoms are pointed in a haphazard fashion in most objects but are all pointed in the same direction in a magnet, first ask the children to think of the metal filings in the plastic bottle or in the "draw-a-picture" toy as atoms. Remind the children that atoms are actually too small to see. Let them observe that the "atoms" are pointing in many directions. Ask a child to stroke the plastic bottle or the plastic covering of the toy in order to turn the metal filings in the same direction. Compare the filings to the atoms in a magnet, all of which are pointed in the same direction.

See if the children are able to use what they have learned from this activity to suggest how they might turn a piece of iron or steel into a magnet. Have a child stroke the needle or nail, always in the same direction, with a magnet 25 to 30 times. Then have them test to see if they have created a magnet by picking up a lightweight metal object with the needle or nail.

Suggested Questions

1. Were the "atoms" able to pull the object (the other child) when they pulled in opposite directions?
2. Were the "atoms" able to pull the object when they pulled together?
3. Before using the magnet to change the direction of the filings, ask, "In what direction are the filings or "atoms" pointed?"
4. After you stroked the filings or "atoms" with the magnet, how were they turned or pointed?
5. Can you think of a way that we could turn a plain piece of iron (or metal) into a magnet?

Experience: Part V

PURPOSE: To make an electromagnet.

MATERIALS: A six-volt battery; about two feet of uncoated or uninsulated wire; a large iron nail; some objects that a magnet will attract.

If the experience on electricity (see pages 64–70) has not been introduced, first establish that electricity comes from a battery and travels through wires. Let the children test in order to see if the nail is magnetic; then help them to wrap the wire around the nail at least 25 times and to attach each end of the wire to a terminal of the battery. Electricity from the battery will pass through the nail to give it a magnetic quality. Let the children then explore the magnetic quality of the nail. The nail will remain magnetized as long as the wires are connected to the battery.

Suggested Questions

1. Is the nail a magnet? Can it pick up metal objects?
2. Now that you have wrapped the wire around the nail and attached it to the battery to get electricity, is the nail a magnet?
3. If we take the electricity (the battery) away, does the nail remain a magnet?

Follow-Up Activities

Discuss the ways in which magnets are used: in motors; as electromagnets used for moving scrapmetal; and in compasses, hearing aids, telephones, and microphones. At home, magnets are used in potholders, to hold pencils to note pads, to hold messages on the refrigerator, to help keep cabinet doors closed, and so on. They are used in toys, such as kissing dolls, magnetic letters, and "draw-a-picture" toys.

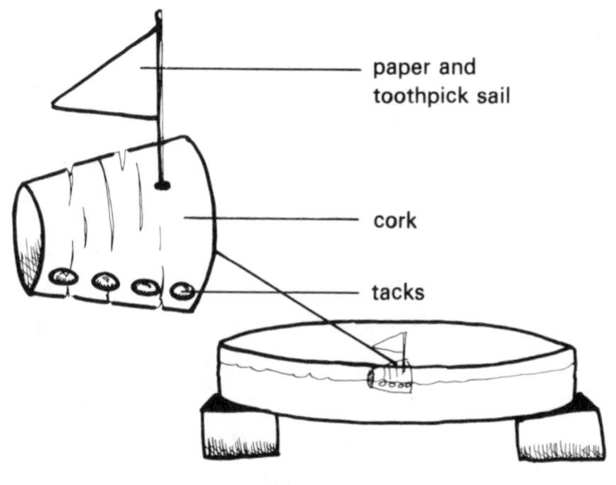

Figure 2-5
Magnetic Sailboat

Let the children make a magnetic sailboat.* Place four or five tacks in the bottom of a cork, and then place a sail made from a toothpick and a triangle of paper in the top of the cork. Float the "boat" in a plastic or glass dish of water that is resting on two blocks or bricks (see Figure 2-5). The children may place a magnet under the dish of water and move the magnet to move the sailboat.

Suggested Readings

Asimov, Isaac. *The Best New Thing.* Cleveland, Ohio: Collins Publishers, 1971.

Branley, Franklin. *Mickey's Magnet.* New York: Scholastic Book Services, 1956.

Freeman, Mae. *The Real Magnet Book.* New York: Scholastic Book Services, 1967.

——. *The Book of Magnets.* New York: Four Winds Press, 1967.

*Adapted by permission of Four Winds Press, A Division of Scholastic Magazines, Inc., from *The Book of Magnets* by Mae Freeman, copyright © 1967 by Mae Freeman.

DINOSAURS

Information for Adults

Dinosaurs were extinct long before the existence of homo sapiens (modern humans). Because there is no written history of dinosaurs, what is known about them today has been interpreted by "reading rocks." As scientists work with rocks, they sometimes discover fossils (petrified remains or imprints of plants and animals that existed long ago). From these fossils, they are able to make calculated guesses about dinosaurs—what they looked like, how much they weighed, how they moved, how they protected themselves, even what they ate.

A fossil print is formed much like a footprint made in mud that dries and hardens and then, over millions of years, changes to rock (becomes petrified). Huge prints of dinosaurs' feet have been found imprinted in rock. Fossils may also be remains of actual parts of animals, such as bones or teeth, or they may even be petrified eggs. Fossils were possibly formed in the following way: When an animal died while watering, it sank to the mud at the bottom of the pond; then, over many years, organic matter in the body was replaced by minerals in the water or mud. The remains that did not decay became petrified.

Dinosaur is a Greek word that means "terrible lizard." However, dinosaurs are not actually lizards, and some dinosaurs were probably not "terrible" but were very gentle. The Age of Reptiles, which is the period when dinosaurs ruled, lasted for more than one-hundred-million years, so not all dinosaurs that have been discovered existed at the same time. There was a great variety of dinosaurs—some with hooves, claws, or webbed feet; some with horns, spikes, or armor. Some dinosaurs had no teeth, whereas others had thousands of teeth; some walked on four feet; some walked upright on two feet. Some dinosaurs are discussed in Appendix I; descriptions of others may be found in the books suggested at the end of this unit.

It is not known exactly why dinosaurs no longer exist, but many reasons have been suggested. Perhaps a combination of the following reasons contributed to the extinction of dinosaurs. (1) Dinosaurs existed when the entire earth had a tropical climate—very warm, with many plants. The climate began to change, and some of the tropical areas became cooler. The cold-blooded dinosaurs could not exist in the cooler climate because they had no protection against the cold and because they were too large to burrow into the ground or find a place to hibernate in order to escape the cold, so they froze to death. (2) In the changes that took place on the earth at this time, some of the lakes and swamps became deserts. There were no longer enough plants for the plant-eating dinosaurs, and they were not intelligent enough to search for new food sources, hence they starved to death. (3) As the plant-eaters died off, there was no longer enough meat for the meat-eaters who fed on the plant-eaters, so they also died. (4) Small mammals came along and ate the eggs of the dinosaurs, so fewer dinosaurs hatched. (5) The mammals that came along were probably more intelligent than the dinosaurs, and the dinosaurs were not able to maintain their place in nature.

Experience: Part I

PURPOSE: To help children to understand how we have learned about dinosaurs, even though they no longer exist.

To make a model of a fossil.

MATERIALS: Clay that may be air dried; small bones, such as a chicken's bones.

If "Experience: Part V" of the unit on "Rocks" has not been introduced, use it at this time as an introduction to dinosaurs. Explain to the children that just as they found new colors, patterns, or crystals when they cracked open rocks, scientists sometimes find animal bones, teeth, or footprints. These are millions of years old and are called fossils. See if they are able to suggest that some of these fossils might be parts of dinosaurs. Pictures from *The National Geographic*, Volume 154, Number 2, August, 1978, pages 152–185 would be excellent to use in this introduction. Be sure to explain that scientists have to learn about dinosaurs from fossils—not only because dinosaurs no longer exist but also because when they did exist, there were no people to leave pictures of what dinosaurs looked like or to tell stories about what they did.

Let the children make fossil models by patting out the clay and making imprints with the bones or their hands. These should be allowed to dry at least overnight to harden. Be sure to explain that real fossils are much older and harder. Discuss where the "clay" that made real fossils came from and how these fossils might have formed.

Suggested Questions

1. What sorts of new things did we find when we cracked the rocks open?
2. Do you know of some animals that lived long ago that we might learn about through studying fossils?
3. How do you think scientists might use fossils to learn about dinosaurs?
4. How do you think a fossil might have been made long ago? (Tell me a story about a fossil.)
5. Are there dinosaurs alive today?
6. Were there any people alive when dinosaurs lived on the earth?

Experience: Part II

PURPOSE: To see pictures of and learn the names for some dinosaurs.

To learn about some dinosaurs' sizes, habits, and nature.

To compare parts of dinosaurs with parts of other animals living today.

MATERIALS: Pictures of dinosaurs found in Figures 2-6a-2-6e. These may be colored in earth tones, or refer to illustrations in "Suggested Reading" for coloring ideas. If desired, the pictures may be prepared for use on a flannel-board.

Figure 2-6a
Tyrannosaurus Rex

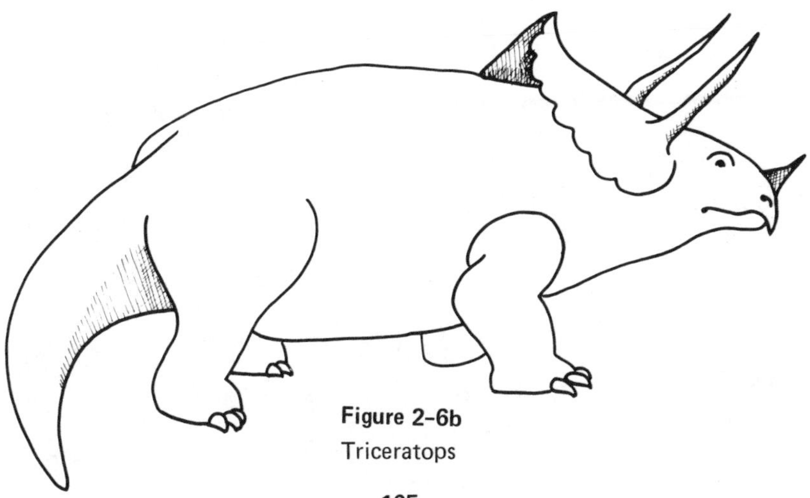

Figure 2-6b
Triceratops

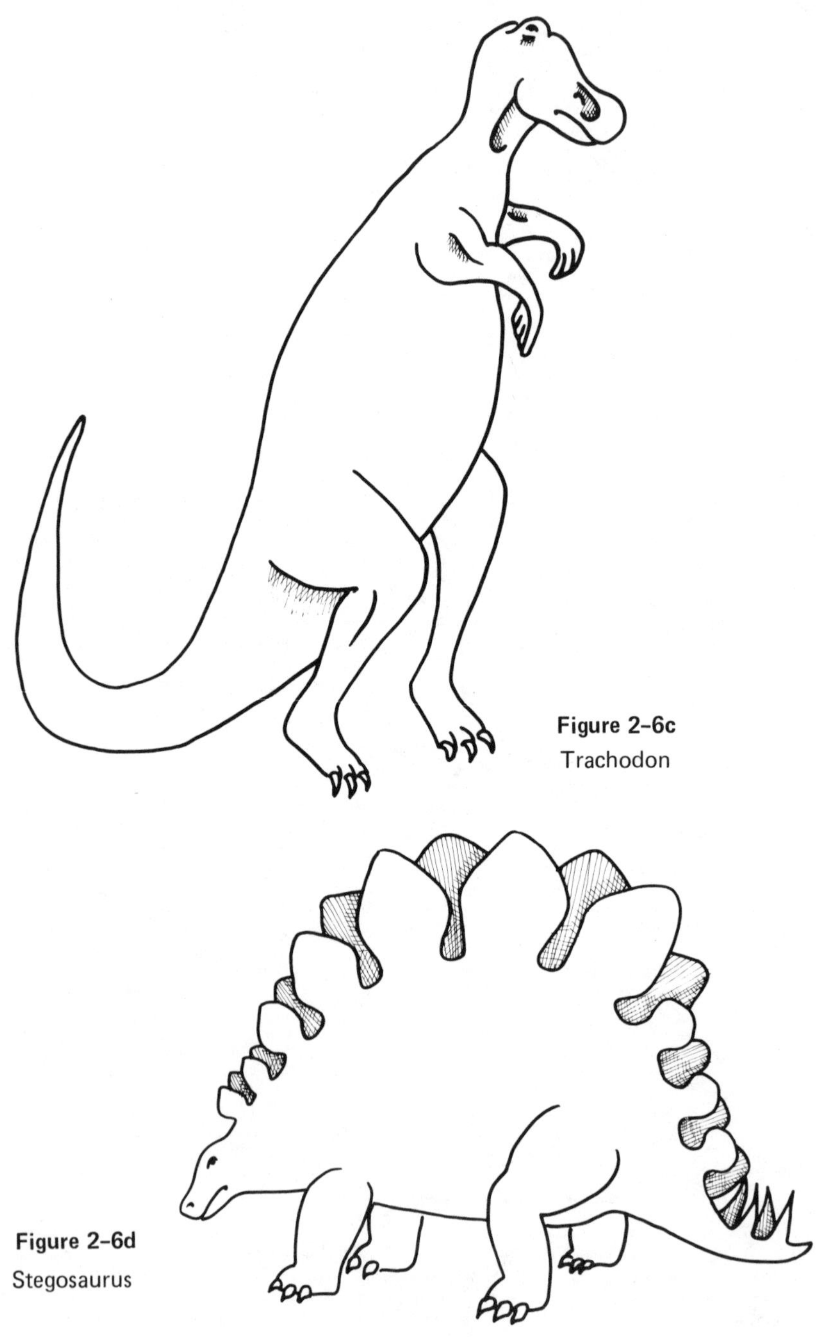

Figure 2-6c
Trachodon

Figure 2-6d
Stegosaurus

Figure 2–6e
Brontosaurus

Show the picture of each dinosaur and give the name for each. Use the information found in Appendix I to discuss characteristics, such as size, weight, how they moved, how they protected themselves, and what they ate.

See if the children are able to pick out the parts of the dinosaurs that resemble parts of animals living today. For example, the Brontosaurus' legs resemble those of an elephant; the Trachodon's front feet resemble those of a duck, and so on. You may wish to provide pictures of some of the animals now living so that the children may make comparisons. Be sure that the children understand that these resemblances do not mean that the dinosaur is related to this animal.

Suggested Questions

1. With each picture of a plant-eater, ask, "Would this dinosaur have eaten people had people lived when dinosaurs were on the earth?"
2. Can you show me a part of this dinosaur that looks like a part of an animal that is living today?

Experience: Part III

PURPOSE: To provide dramatic play experience with dinosaurs.

MATERIALS: Plastic dinosaur models; a dinosaur village. This may be made with the children's help. You'll need papier-mâché, a large piece of styrofoam, small pieces from artificial greenery to be used as plants and trees, small sections of tree limbs to use as logs, small rocks, blue cellophane to be used as water (a blue transparent report cover may be used).

Help the children to build a dinosaur village, forming caves for dinosaurs to hide in with papier-mâché and painting them brown. A swamp may be made by carving a pond shape about one-quarter of an inch deep into the styrofoam. Cut the blue transparent report cover to the shape of the pond and place it in the carved-out area. Add rocks and "logs"; then add the artificial greenery (attach by forcing one end into the styrofoam).

Let the children match the plastic dinosaur models to the pictures used in the previous experience and again discuss the names and habits of the dinosaurs. Allow the children to use the plastic models and the dinosaur village for dramatic play.

Suggested Questions

1. What is the name for the dinosaur you are playing with?
2. Do you think Tyrannosaurus Rex would have been a good swimmer—why or why not?
3. What does that dinosaur like to eat?
4. If Tyrannosaurus Rex attacked Triceratops, how would Triceratops protect himself? How would Stegosaurus protect himself? Trachodon? The others?

Follow-Up Activities

Make "Dinosaur Bread," a whole wheat bread shaped like a dinosaur, from the following recipe:

1-1/4 cups warm water 1 package dry yeast 2 Tablespoons brown sugar	Add the sugar and the yeast to water and let stand five minutes.
3 cups whole wheat flour 2 Tablespoons cocoa 1/2 Tablespoon salt 2 Tablespoons molasses 2 Tablespoons oil	Mix the dry ingredients and add the two remaining ingredients.

In a large mixing bowl, mix all ingredients with a spoon. When you can no longer stir the mixture, turn it onto a floured board or table and knead for 10 minutes. (White flour may be kneaded into the dough to make it less sticky.) Roll the dough into a large ball, lightly coat it with cooking oil, and return it to the mixing bowl. Cover the bowl with a cloth and place it in a warmish location (70° F. or warmer). Allow the dough to rise for one hour, less if necessary, or overnight if necessary. Flour hands and shape the dough by patting out pieces for the head, body, legs, tail, and large spikes. The illustration in Figure 2-7 suggests a Stegosaurus shape. Attach the smaller pieces to the body pieces by moistening (with a finger tip dipped in water) the edges to be joined. After the dinosaur has been shaped, raisins may be placed in the uncooked dough to represent eyes, nose, and toes; sliced almonds become additional bony plates. Brush

Figure 2-7

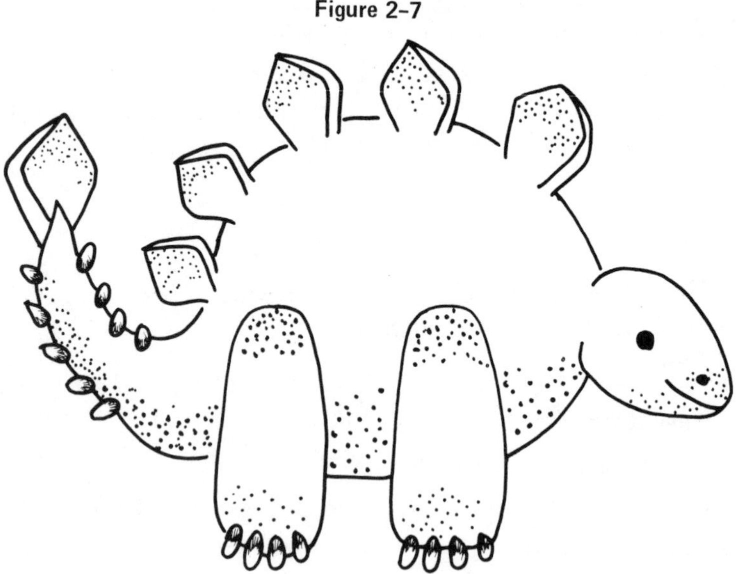

Note: The idea for "Dinosaur Bread" was taken from the book, *Bread Sculpture: The Edible Art*, by Anne Wiseman, San Francisco, Calif.: 101 Productions, 1975.

with beaten egg and bake on an ungreased cookie sheet at 350° F. for 30 to 45 minutes.

Have the children design a twentieth-century dinosaur. If dinosaurs could come back to live on the earth, what would they be like? How large would they be; what color would they be; where would they live; and what would they eat?

For a math experience, convert the dinosaur pictures into dot-to-dot drawings. Cover the dinosaur pictures one at a time with tracing paper and "outline" each dinosaur with dots placed about one-half-inch apart. Number the dots consecutively beginning with "1" and let the children complete the outline of the dinosaur by drawing lines between each dot, beginning with dot number "1," moving on to dot number "2," then to "3," and so on.

"Fossil" prints may be used as gifts. Use the child's hand as the imprint and make a hole in the top of the clay while it is wet. Shellac the dry fossil and run a ribbon through the hole for hanging.

Suggested Readings

Aliki. *My Visit to the Dinosaurs.* New York: Thomas Y. Crowell, 1969.

Childcraft. *The How and Why Library: Prehistoric Animals.* San Francisco, Calif.: Field Enterprises, 1976.

Cole, Joanna. *Dinosaur Story.* New York: Scholastic Book Services, 1974.

Craig, Jean. *Dinosaurs and More Dinosaurs.* New York: Four Winds Press, 1965.

VOLCANOES

Information for Adults

The origin of a volcanic mountain is quite different from that of an ordinary mountain. A volcano is formed when hot molten rock, steam, and gas force their way from the center of the earth to the surface. This material travels through a crack that reaches from the hot molten rock to a point on or near the surface of the earth. The molten material, which is called magma, cools and hardens when it reaches the surface and is then called lava. Sometimes a very porous rock called pumice is formed if there is an abundance of gas in the magma.

As lava builds up around the opening in the earth, a cone-shaped volcanic mountain is formed. An opening called a crater remains in the center of the volcano. This opening is usually the original hole through which the magma began to flow. A mountain that did not originate in this way will never become a volcano.

Some volcanoes were formed millions of years ago, and although most of these are now inactive, many active volcanoes still exist. Most of the very active volcanoes existed before there were modern humans (homo sapiens) on the earth.

A volcano does send out a warning before it erupts. The ground nearby may shift (like a small earthquake) for several days; steam may begin rising from the crater; and there may be rumbling noises that come from the inside of the volcano. Finally, the volcano will swell just before it erupts. For the children, the eruption of a volcano may be compared to a bottle of pop that builds up pressure inside if it is shaken, so that it blows its top. The pop, including liquid and gases, spews out the top and over the sides of the bottle.

Experience: Part I

PURPOSE: To demonstrate how a volcano looks when it erupts.

MATERIALS: A volcano model (made of wire and papier-mâché); a juice glass filled with vinegar colored red with food coloring; two tablespoons of baking soda.

If the children have not asked specific questions about

volcanoes, discuss with them their ideas of what a volcano is and what it does. If they have not seen a volcano on television, in a movie, or in a book, show them a picture of a volcano at this time. The books in "Suggested Readings" at the end of this unit offer a source of pictures.

Help the children to build a model volcano by forming stiff wire into a cone shape with a hole in the point of the cone. Cover the cone with papier-mâché, allow it to dry, and then paint it with brown or grey tempera to resemble a mountain. (In order to prevent the tempera from chipping, powdered tempera may be mixed with liquid floor wax instead of water, or water base tempera, when dried on the surface of the cone, may be covered with shellac.) See Figures 2-8a through 2-8c.

Place the glass containing the red vinegar inside the hole of the cone. It may be necessary for the glass to rest on a block of wood in order to make it the correct height. Explain to the children that the hole in the center of a volcano is called a crater.

When one tablespoon of baking soda is added to the glass of vinegar, bubbles and gases will form on the surface, and the red vinegar will flow over the sides of the volcano. This gives the appearance of an erupting volcano. The second tablespoon of soda may be added to the vinegar remaining in the glass in order to erupt the volcano a second time.

Figure 2-8a

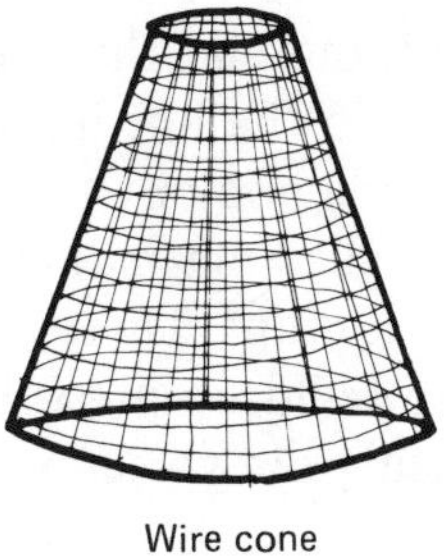

Wire cone

Figure 2-8b

Cone covered with papier mâché

Figure 2-8c

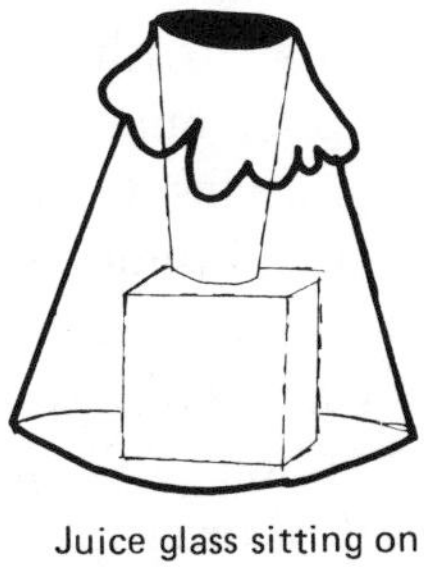

Juice glass sitting on block inside cone

Suggested Questions

1. What is a volcano? What does it look like?
2. Is there a hole inside our model volcano? What is the hole in a real volcano called?
3. What do you see forming at the opening of the volcano?
4. What happened when the gases and bubbles built up inside the volcano?

Experience: Part II

PURPOSE: To understand, in part, what causes a volcano to erupt.

MATERIALS: A bottle of pop or plain bottled soda water; an empty glass.

To begin to demonstrate what causes a volcano to erupt, first pour some of the soda pop or soda water into a glass in order to observe the bubbles and gases on the surface and in the liquid. Place one thumb over the bottle of pop or soda water and shake it vigorously. Observe the bubbles and gases inside the bottle that build up and then spew, along with the liquid, out the top of the bottle and down the sides when you remove your thumb.

Explain that the opening in the top of the bottle may be compared to the opening in the center of a volcano and that the soda pop, bubbles, and gases may be compared to the hot liquid and gases inside the earth. Clarify that it is not soda pop or vinegar and soda inside a volcano but hot melted rock (called magma) and that no one shakes a volcano in order to make it erupt. There are gases and bubbles in the liquid rock, and these build up pressure or "bubble up" (partly because of heat), causing the volcano to overflow or erupt.

Suggested Questions

1. What is a volcano? Are there any volcanoes near where we live?
2. What do you see inside the glass and just on the surface of the liquid when I pour some of the soda pop into the glass?
3. What is happening as I shake the liquid inside the bottle?
4. What happens when many bubbles and gases build up inside the bottle?
5. Is there an opening through which the liquid and the gases and bubbles may escape or leave the bottle?
6. How is this activity like a real volcano erupting?

Follow-Up Activities

The adult may wish to discuss with the children that, in the beginning of the formation of the earth, volcanoes were helpful because they gave off carbon dioxide, which plants needed to grow, and because they gave off water vapor, which changed into rain. Volcanoes were possibly people's first source of fire. Volcanoes have helped to form fossils because plants and animals were covered with the liquid magma, which then hardened and preserved the prints.

The children may make volcanoes in the sand box. Give them plastic or paper cups to place inside their cone-shaped sand mountain. These may be filled with red vinegar and baking soda may be added to create the eruption.

Suggested Readings

Allen, Hazel. *Up From the Sea Came an Island.* New York: Scribner's, 1962.

Ladyman, Phyllis. *Inside the Earth.* Brokhampton Press Ltd., Reading, Mass.: Addison-Wesley, 1969.

Lauber, Patricia. *Junior Science Book of Volcanoes.* Champaign, Ill.: Garrard Publishing Co., 1965.

May, Julian. *Why the Earth Quakes.* New York: Holiday House, 1969.

Parker, Bertha Morris. *The Golden Book of Science.* New York: Golden Press, 1963. See "Mountains That Smoke" p. 55.

Stephens, William. *Islands.* New York: Holiday House, 1974.

Note: Although none of these books deals exclusively with volcanoes, parts may be adapted for the study of volcanoes with young children.

BIRDS

Information for Adults

This unit may be presented at any time of the year, but it is most appropriate for winter and early spring. There are few other animals that may be observed during the winter, and this is a time when birds may need some assistance in finding an adequate food supply. In the early spring, birds may be observed migrating to their summer homes and building nests.

Birds are distinguished from other animals in several ways: The most obvious of these is that birds have feathers and that they are the only animals that do. Birds have five basic types of feathers: (1) filoplumes, which have an almost invisible shaft running through the center; (2) bristles, which are modified feathers found on some birds in the mouth and nostril area; (3) contour feathers, which form the bird's outline and help to give it characteristic color markings; (4) semi-plumes, which are fluffy feathers with a central vane; and (5) down feathers, which are fluffy feathers lacking a central vane. The latter three types of feathers are of more interest to children, as they may be thought of as a bird's clothes.

Contour feathers, which are the outer feathers, offer some

protection from water and may be compared to an overcoat. These feathers have oil on them to help repel water, and their density helps keep water from reaching the bird's skin. Semi-plumes, which may be found just under the contour feathers, may be compared to outer clothing, and down feathers, found next to the bird's body, may be compared to underwear. Both semi-plumes and down offer insulation and warmth.

Colors and patterns of feathers may be used in order to help distinguish one bird from another. An aid to help children remember the color characteristics of each type of bird is to divide the parts of the bird into blocks of color representing the head, the back, the breast, and the tail. For example, a robin would be distinguished in the following manner:

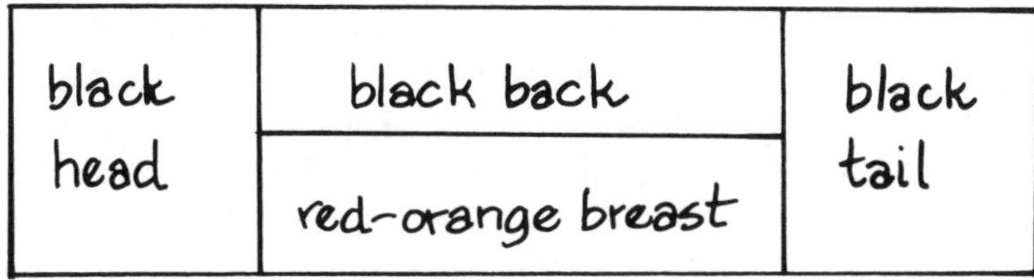

Figure 2-9

Different kinds of birds may be spotted in one area at different times of the year. Because of bird's habits of migration, one area may have as many as four different bird populations during the year—summer residents, winter residents, permanent residents, and migrants. Birds probably migrate to find an adequate source of food during the winter as well as to find protection from the cold. Most scientists believe that familiar sights on the ground, such as large water surfaces (rivers and lakes), help birds to find their way as they migrate. These migrations may cover several thousand miles or only a short distance and can be hazardous for the birds: They may meet obstacles, such as bad weather, hunters, or even large windows to crash into. A cutout pattern that resembles a diving hawk has been designed at the Smithsonian Institution. If placed on large windows, it should frighten birds away and thus protect them from crashing into the window. To order the pattern, write to the following address: Smithsonian Institution Mail Order Division, 7955 Angus Court, Springfield, Virginia 22153.

The main way to attract birds for observation is to provide them with food. Different kinds of birds eat different kinds of food, so it may be desirable to provide a variety of food in order to attract a variety of birds. Some good foods to offer winter birds include wild bird seed, sunflower seed, cracked corn, bread crumbs, and berries. In an area safe from cats, the food may be scattered on the ground, since some birds prefer to find their food in more natural surroundings. Other birds are attracted to feeders, which may be easily constructed or purchased already made.

When observing birds, notice their beaks and how they use them. A bird has a tongue but no teeth, and it uses its beak for breaking up food. Each bird has a beak shaped to aid it in eating the kind of food that it prefers. Most birds have four-toed feet adapted for their habits. For instance, a duck's foot is webbed for swimming; a woodpecker's foot is adapted for climbing and clinging to the sides of trees; a cardinal or sparrow has a perching foot; a heron or crane has a wading foot; and a chicken has a walking foot (see Figures 2-12a-e).

Birds build nests in the spring and summer—not really as homes, but for raising young. Fall and winter are the only times of the year that birds' nests should be collected. Each bird builds its own particular style of nest. The size of the nest and the materials with which it is built will vary according to the variety of the bird that built it. Some materials birds commonly use to build nests include straw, dry grass, twigs, pine needles, small sticks, hairlike fibers, thread, and string. Some nests may be elaborate in design, woven together with hair or even with spider webs or held together with mud or spittle, whereas other nests may be merely piles of grass, sticks, or straw.

Birds may be distinguished one from the other by size and shape. The three basic shapes are chunky (like robins, blackbirds, blue jays, cardinals, owls, and woodpeckers), plump or round (like chickens, doves, and bobwhites), and slender (like sparrows, chickadees, swallows, hummingbirds, and finches). The three common sizes are sparrow size (5″), robin size (7″), and crow size (14″). The pattern shown in Figure 2-10 may be adjusted to these three sizes and cut from cardboard so that the children may use them for comparison as they observe and identify birds.

Birds may also be distinguished by the sounds they make as well as by their activities or habits. The sounds that birds make may range from a beautiful song to a screech, a tap to a twitter. Birds such as killdeer, whip-poor-wills, or bobwhites call their own names. Crows caw, jays swalk, some owls whoot, and some sparrows chirp—"chip, chip, ip-ip-ip-ip-ippp". Some warblers slur something that sounds like "witchity, whitchity," ovenbirds call "teacher teacher," and towhees call, "Drink your tea."

Activities or habits of birds include behavior, methods of locomotion, and daily routine. Birds spend their time on the ground, perched in trees, in flight, or on the water. Some birds walk with a strut; others take short little steps, bobbing their heads as they go. Some birds waddle as they walk, and some long-legged birds appear to teeter as if they're going to fall. Most, though not all birds fly and may be distinguished one from the other by their patterns of flight. Some birds soar; some hover like a helicopter; some flutter as they fly; some flap their wings in an even pattern and fly in a straight line; others rise and fall as if riding a wave.

Figure 2-10

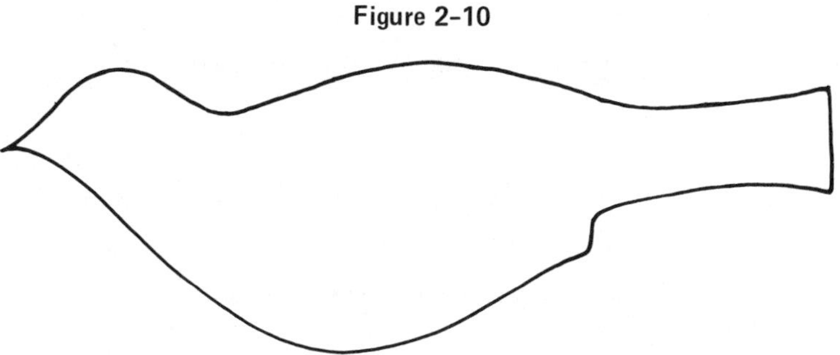

Experience: Part I

PURPOSE: To begin to identify some of the characteristics of various birds.

MATERIALS: A bird feeder or food to attract birds; cardboard shapes representing the three basic sizes of birds (see "Information for Adults" and Figure 2-10); blocks of color to represent common birds in your area (see "Information for Adults" and Figure 2-9).

Discuss with the children the reasons birds may have difficulty getting enough food during the winter. (For warmth, insects and earthworms are hidden deep under the ground, under rocks, or under bark, and berries and seeds are gone or buried under snow.) Help the children set up a feeding station (preferably outside a window) in order to attract birds.

Provide cardboard shapes that represent the three basic sizes of birds and let the children use these in order to help determine the size of the birds that they observe. You may wish to speak of these sizes in terms of small, medium, and large.

Help the children begin to identify birds by color and pattern by using the "block" technique suggested in "Information for Adults."

While observing birds, discuss with the children some observable habits, methods of locomotion, and so on.

Suggested Questions

1. What size is this bird you're watching—large (14″), medium (7″), or small (5″)?
2. What are the main colors you see on this bird and where are these colors (what color is the head, back, tail, and breast)?
3. Where does this bird seem to spend most of its time?
4. How does this bird walk? Can you show me?

5. Have you seen this bird fly? How did it fly?
6. Have you heard this bird make a sound? What kind of sound?

Experience: Part II

PURPOSE: To understand that birds' beaks are adapted for the type of food they eat.

To understand that birds' feet are adapted for their habits or lifestyles.

MATERIALS: Pictures of birds in Figures 2–11a–f (colored and prepared for a flannelboard if desired); pictures of birds' feet in Figures 2–12a–12e (colored and prepared for the flannelboard if desired); pictures of birds that illustrate some examples of different types of beaks or feet (*Ranger Rick's Nature Magazine*, published by the National Wildlife Federation, is an excellent source of bird pictures).

Show the pictures illustrating the various types of birds' beaks and discuss what kind of food each bird eats and how its beak is adapted to help it eat this type of food. Display a variety of pictures of birds and let the children identify the type of beak each has and suggest the type of food each would eat. Let them categorize the bird pictures according to the beaks. See if the children are able to identify some of the different types of beaks on the birds observed at the feeding station.

Show the pictures illustrating the various types of birds' feet and discuss what sort of lifestyle each type of foot is adapted for. Display a variety of pictures of birds and let the children identify the type of foot each has, and then suggest this bird's lifestyle. Let them categorize the bird pictures according to the types of feet the birds have. Suggest that the children observe the feet of the birds at the feeding station in order to determine each bird's lifestyle.

Figure 2-11a

HAWK: Uses its hooked beak for tearing the flesh of small animals.

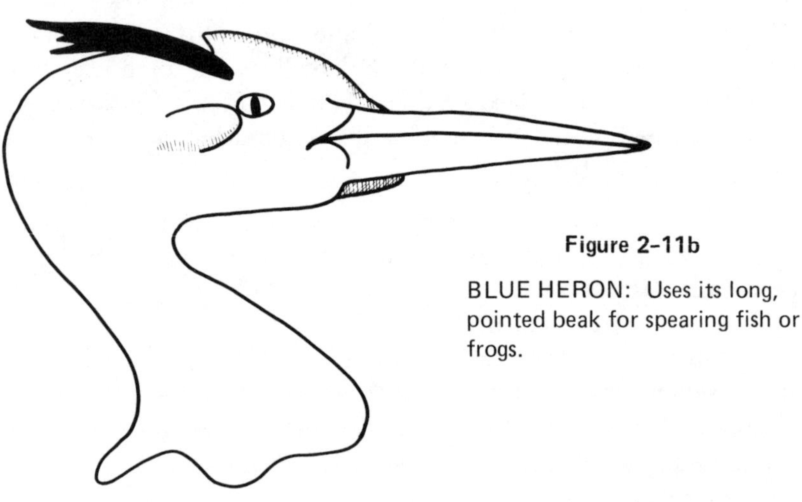

Figure 2-11b

BLUE HERON: Uses its long, pointed beak for spearing fish or frogs.

Figure 2-11c

ROBIN: Uses its short but pointed beak for digging insects and worms.

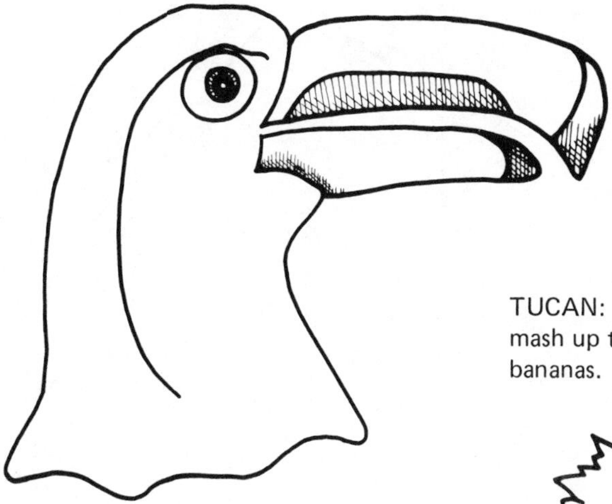

Figure 2-11d

TUCAN: Uses its large beak to mash up tropical fruits, such as bananas.

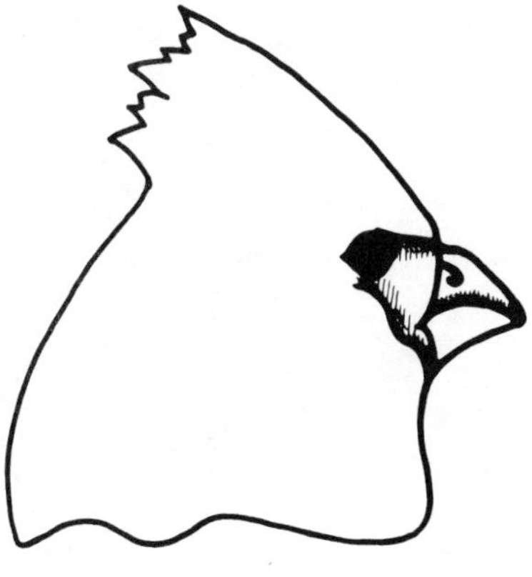

Figure 2-11e

CARDINAL: Uses its short, thick, cone-shaped beak for cracking seeds and mashing berries.

Figure 2-11f

WOOD DUCK: Uses its flat scoop-like beak to scoop fish from the water.

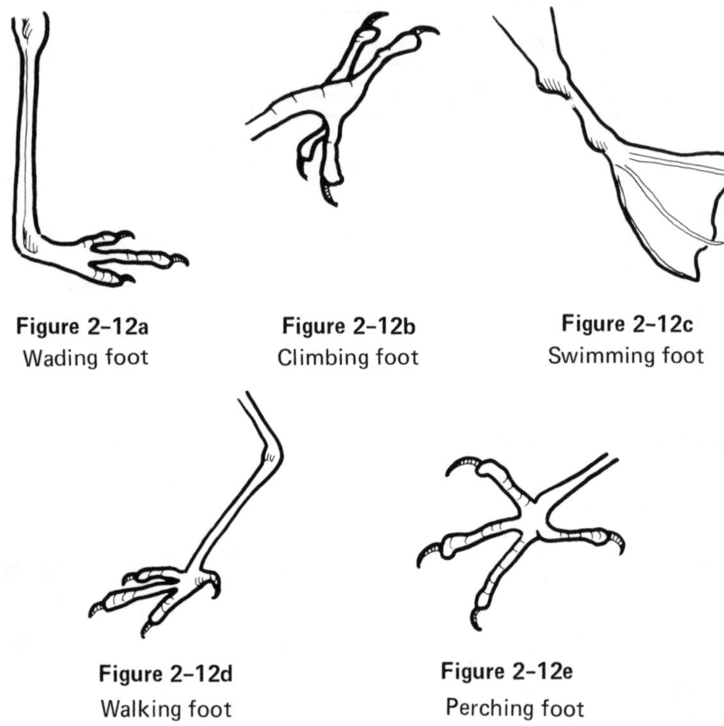

Figure 2–12a Wading foot

Figure 2–12b Climbing foot

Figure 2–12c Swimming foot

Figure 2–12d Walking foot

Figure 2–12e Perching foot

Birds: Feet

Suggested Questions

1. What sort of beak does this bird have (short and thick, short and pointed, long and slender, flat, hooked, or large and curved)?
2. What type of food is this bird eating?
3. What type of foot does this bird have (a walking foot, a perching foot, a swimming foot, a wading foot, or a climbing foot)?
4. Where does this bird spend most of its time (in the water, perched in a tree, in flight, or on the ground)?

5. From these pictures, can you find a bird that spends most of its time in a tree, on the ground, or in the water?

Experience: Part III

PURPOSE: To explore the differences in feathers of different types of birds.

To begin to learn what purpose different feathers serve and how to identify different types of feathers.

MATERIALS: A collection of feathers from different types of birds—and a tail feather, wing feather, neck feather, and breast feather (all from one type of bird if possible, perhaps a domestic bird, such as a chicken or turkey); a piece of poster board.

Discuss with the children some of the different types of feathers that birds have and the purpose of each type of feather. Specifically identify a countour, a semi-plume, and a down feather and discuss how these may be compared to clothes. Refer to "Information for Adults" for details.

Let the children categorize the feathers into three groups—contour, semi-plumes, and down.

Make a chart that identifies a tail feather, a wing feather, a neck feather, and a breast feather by mounting each of these feathers on poster board along with a picture of the bird from which they came. Draw a line from each feather to the part of the bird from which it came. The children may use the chart to determine from what parts of the bird other feathers have come. Provide the children with a collection of feathers or have them collect feathers to classify according to the part of the bird from which they came. This may be done by matching the collected feathers to those on the chart.

Suggested Questions

1. What type of feather is this (contour, semi-plume, or down)?
2. What purpose does this feather serve, and what piece of

clothing may it be compared to (overcoat, outer clothing, or underwear)?

3. From what part of the bird did this feather come?
4. Do most birds have tail feathers, wing feathers, breast feathers, and neck feathers?
5. Are all birds able to fly well?

Experience: Part IV

PURPOSE: To see some of the materials that birds use to build nests.

To gain an appreciation for the bird's ability to build a nest.

MATERIALS: Old birds' nests that have been sprayed with insecticide.

Provide the children with an old bird's nest and let them examine it. Ask them if they think they could build a bird's nest. Then allow them to take the old nest apart in order to discover all the materials that the bird used and let them try to put the nest back together again. Be sure to have the children wash their hands well after this activity. Discuss the difficulty of the task of building a nest.

See if the children are able to think of some ways they can help birds in their nest building, such as providing them with materials. Let the children scatter straw and grass on the ground and hang string, small strips of fabric, and cotton filaments from cotton balls in the trees.

Suggested Questions

1. What did the bird use to build this nest?
2. How did the bird hold the materials together?
3. What size is this nest? Do you think it belonged to a large, medium, or small bird?

4. What could we do (or provide) in order to help birds build their nests?
5. What sort of materials should we provide?

Follow-Up Activities

Go on a bird identification walk in order to see how many different varieties of birds may be spotted. A book, such as *Birds: A Guide to the Most Familiar American Birds*, by Herbert S. Zim and Ira N. Grabielson (a Golden Nature Guide Series), will be very useful in bird identification.

Let the children make a simple but effective bird feeder from a pine cone. Tie a string for hanging purposes at the smaller end of the pinecone, and then spread the cone with a mixture of peanut butter and cornmeal (approximately two parts of peanut butter to one part of cornmeal). Roll the coated pine cone in wild bird seed (this may be purchased from most grocery stores) and hang it from a tree limb.

Let the children make "bird seed cakes" from suet (hard fat) or lard, wild bird seed, squares of loose textured netting or mesh bags (such as those containing oranges, onions, or potatoes), and pieces of string (about one foot long). Melt the suet or lard (about one cup), allow it to harden, and then melt it again. While melted the second time, mix in about two (or more) cups of wild bird seed. Allow this mixture to harden slightly, and then place about three tablespoons into the center of each net or mesh square (the pieces should be about six inches square). Gather all four corners of the square and tie with a string so that the "cakes" resemble sachet bags. Hang from a tree limb with the string.

Suggested Readings

Dolch, Edward, and Dolch, Marguerite. *Friendly Birds.* Champaign, Ill.: Garrard Publishing Co., 1959.

Earle, Olive L. *Robins in the Garden.* New York: Morrow, 1953.

Eastman, P. D. *The Bird Nest.* New York: Random House, 1968.

Farra, Richard. *The Hungry Snowbird.* New York: Coward, McCann & Geoghegan, 1975.

Grans, Roma. *It's Nesting Time.* New York: Thomas Y. Crowell, 1964.

——*Bird Talk.* New York: Thomas Y. Crowell, 1971.

Georgiou, Constatine. *The Nest.* Irvington, N.Y.: Harvey House, Inc., 1972.

Holl, Adelaide. *The Remarkable Egg.* New York: Lothrop, Lee and Shepard Co., 1965.

Selsam, Millicent, and Hunt, Joyce. *A First Look at Birds.* New York: Scholastic Book Services, 1973.

Skorpen, Liessel Moak. *Bird.* New York: Harper & Row, Pub., 1976.

Stiles, Bennett. *Dougal Looks for Birds.* New York: Four Winds Press, 1972.

Wildsmith, Brian. *Birds.* New York: Franklin Watts, Inc., 1967.

——*The Owl and the Woodpecker.* New York: Franklin Watts, Inc., 1972.

Zim, Herbert S., and Grabrielson, Ira N. *Birds: A Guide to the Most Familiar American Birds.* New York: Golden Press, 1956.

3
SPRING

WEATHER MEASUREMENT

Information for Adults

Wind, which is nothing more than moving air, is essentially responsible for changing the weather. As winds move into and out of a geographic area, they can bring either cool, warm, moist, or dry air. Winds result from the unequal heating of land and water surfaces on the earth. The movement of air is created by one mass of air (usually warmer air) rising and another mass of air (cooler air) moving in to take its place. Air rises for one of two reasons. Either it has become warmer, less dense, and thus lighter; or it is forced to rise, either over a mass of cooler, denser air that stays near the ground or over some massive solid object, such as a chain of mountains.

Wind is measured in two ways: direction (determined by a wind or weather vane) and speed (determined by an anemometer). A weather vane is nothing more than a pointer, like an arrow, with one smaller surface, the head, and one larger surface, the tail. As wind blows, it exerts more force on the larger or more massive surface and swings the vane to point in the direction from which the wind is coming. There are stationary markers attached to the vane to tell which direction the vane is pointing. The direction from which the wind is coming is important, because it can indicate the type of weather that may be expected. For example, north winds bring cooler weather, and south winds bring warmer weather. In the southern hemisphere, the opposite would be true.

An anemometer consists of three or four cups mounted on a pole so that they may turn freely. The cups catch the wind, causing them to spin around, and the faster the wind is blowing, the faster they spin. The anemometer is usually wired to an instrument that records the speed of the wind. The speed of the wind can indicate rapidly moving winds created by severe weather, such as gales, storms, hurricanes, and tornadoes.

Moving air or wind can cause changes in weather from warm air to cold air and vice versa. These temperature changes are measured by a thermometer, a slender glass tube that amplifies as much as possible small changes in the volume of a substance. The changes in volume are created by changes in temperature. In a thermometer that measures body temperature, the substance that changes volume is usually mercury, whereas in a weather or air thermometer, the substance used is usually alcohol that has been colored red.

Wind may also bring moisture in the form of rain. The instrument used to measure the amount of rainfall is a rain gauge. This consists of a container that collects rainfall, so that the depth of the accumulation of water may be measured in inches or centimeters. A rain gauge used for official measurement includes an instrument that can electronically record the amount of rainfall.

Experience: Part I

PURPOSE: To discover that wind is moving air.

MATERIALS: Pieces of cardboard (one for each child) that may be used for fanning.

Ask the children whether or not they feel any wind in the room. See if they are able to suggest a way that the air in the room could be moved about (perhaps by fanning with the cardboard or by another method that they desire). While the children are moving the air in the room, ask whether they now feel any wind.

Suggested Questions

1. Is there any wind in the room right now?
2. How could we move the air in this room?
3. Do you feel wind when you are moving the air?
4. How is wind made? What is wind?

Experience: Part II

PURPOSE: To learn how the direction of the wind may be determined by using a weather vane.

MATERIALS: A weather vane made from one unsharpened pencil, one plastic drinking straw, one straight pin, a piece of cardboard; a stapler, a map of your country (optional).

Cut a pointer and tail for the weather vane from cardboard (each piece should be a double thickness). Staple these securely to each end of a drinking straw. Stick the straight pin through the middle of the straw and into the center of the pencil eraser (see Figure 3-1).

Discuss with the children what is meant by the directions north, south, east, and west and how wind from each of these directions may bring different types of weather. A map will be helpful in this discussion.

Outside, help the children to mark north, south, east, and west on the ground. Let one child stand in the center of these markings and hold up the weather vane so that it will point in the direction from which the wind is coming.

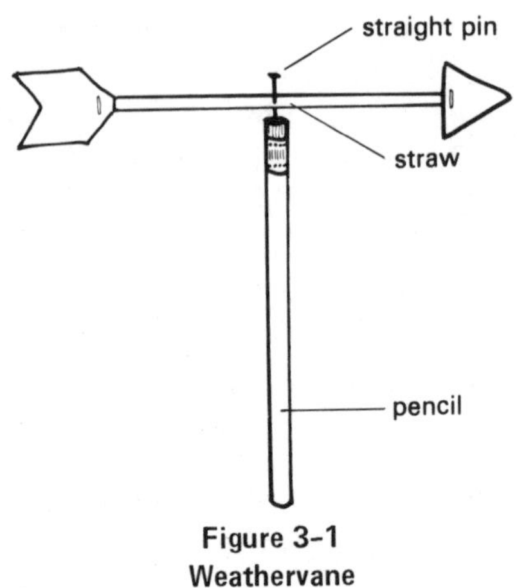

Figure 3-1
Weathervane

Suggested Questions

1. In which direction is the weather vane (arrow) pointing?
2. If it points in the direction from which the wind is coming, then what is the direction that the wind is coming from?

Experience: Part III

PURPOSE: To learn how the speed of wind may be measured by an anemometer.

MATERIALS: An anemometer made from one dowel (approximately 1/2″ to 1″ in diameter and 3 feet long), one headless finishing nail 2 1/2 inches or longer, a hammer, two strips of lattice or lightweight wood, such as balsa (each measuring approximately 3/4″ X 1/8″ X 14″), four plastic or paper cups (three of one color, one of a different color—perhaps red), the plastic cover to a ball point pen, glue.

Glue the strips of wood together in the center to make a cross and bore a hole in the exact center of the cross. Remove the clip or extension on the ball point pen cover and force the cover

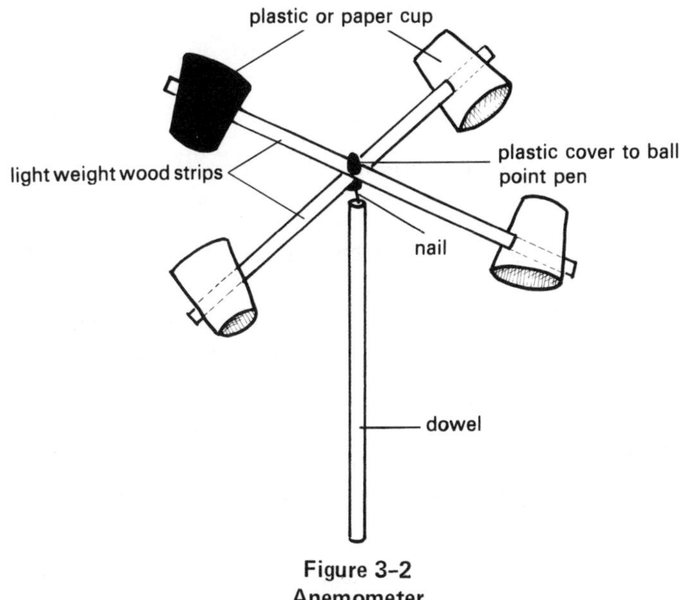

Figure 3–2
Anemometer

into the hole in the center of the cross. Cut two slits on the opposite sides of each cup and slip each end of the cross into the slits of a cup (see Figure 3-2). Hammer the finishing nail into one end of the dowel and balance the cross on the nail with the ball point pen cover by placing the ball point pen cover over the nail.

Outside, in an open area, bury the dowel end of the anemometer a few inches into the ground so that it can stand on its own or have a child hold the anemometer upright in order to catch the wind. Let the children help count the number of times the red cup goes around in two and one-half minutes. This will give the approximate speed of the wind in miles per hour. Discuss with the children the significance of knowing the wind speed.

Suggested Questions

1. How many times did the red cup go around as we counted?
2. How fast is the wind blowing?

3. Why is it helpful to know the speed of the wind?
4. Could wind speed have anything to do with the type of weather we have?

Experience: Part IV

PURPOSE: To help children understand how to use a thermometer.

To help children understand how a thermometer works.

MATERIALS: A room thermometer large enough for the red line to be clearly visible; two bowls (one filled with ice water; the other filled with warm water); a *long*, red balloon; a slender straight-sided jar; enough red-colored water to fill the jar one-quarter to one-half full; a marker pen; a candle; matches.

Let the children feel the water in both bowls in order to establish that one bowl of water is warm and the other is cold. Let them place the thermometer first in the cold water and then in the warm water in order to see what happens to the red line of the thermometer (whether it goes up or down). Let the children test the thermometer in warm and cold air in order to see that it works the same as in the water.

A simple way to demonstrate what makes a thermometer work is to discuss the movement of the red line in terms of expansion and contraction. Have a child make a balloon expand, or get bigger, and contract, or get smaller. Now place the balloon inside the slender jar and have the child make the balloon expand and contract inside the jar in order to see that as it gets larger, it fills more of the jar and that as it gets smaller, it fills less of the jar. Compare this to the liquid (usually alcohol) expanding and contracting inside the tube of the thermometer to fill more or less of the tube.

The children may make a kind of homemade thermometer by filling the slender jar (or a test tube, if available) one-quarter to one-half full of water that has been colored red. Alcohol should not be used in this experiment, since the thermometer is used with an open flame and could ignite very easily. Mark the level of the water with the marker. Then hold the jar or test tube over a lighted candle in order to heat the water. As it expands, the level of the water will go up or rise slightly above the mark. The more slender the tube, the more obvious will be the expansion. It may be desirable to discuss with the children that the tube of a thermometer is so slender that it amplifies, or helps us to see more closely, the expansion of the liquid.

Suggested Questions

1. What happens to the red line when you put the thermometer in the warm water? In the cold water?
2. What happens to the red line of a thermometer as the weather or air gets warmer? Colder?
3. What does the balloon do in the jar as it expands? Does it fill more of the jar?
4. Can you see as much of the red balloon in the jar when the balloon contracts?
5. What does the red line (alcohol) do inside the tube of the thermometer when it expands? Does the red line (alcohol) fill the tube as it expands? What happens as the alcohol contracts?
6. What happened to the red water when it was heated? Did it go up or fill more of the jar or test tube?

Experience: Part V

PURPOSE: To learn how rainfall is measured.

MATERIALS: A tall, slender, straight-sided jar, such as an olive jar; a wide-mouth, straight-sided jar, such as a peanut butter jar; a ruler; a funnel.

Help the children pour one inch or centimeter of water into the peanut butter jar. Then pour the water into the olive jar and mark off this amount into ten equal parts. The depth of water in tenths of an inch will be much easier to read in the slender olive jar.

Outside, put the olive jar on a level surface in the open and put the funnel in the mouth of the jar. After it rains, read the amount of rain caught in tenths of an inch or centimeters, as marked on the jar.

Since the children will not have had much experience with fractions, it may be necessary to cut a piece of paper into a one inch piece and then into ten equal parts and to let the children fit the correct number of pieces together to match the depth of rainfall. For instance, if 4/10″ of rain fell, the children would put four pieces of paper together in order to "see" the amount of rainfall.

Suggested Questions

1. How much rain did our rain gauge catch? How many marks (tenths of an inch) deep is the water?
2. How many parts of an inch will you put together to see how deep the rain was?

Follow-Up Activities

Let each child make a variation of a weather vane with a dowel or plastic straw and a strip of crepe paper fifteen to twenty inches long. Tape or staple one end of the crepe paper to one end of the dowel or straw. The wind will blow the tail (the loose part of the paper) in the direction it is blowing, so that the other end of the crepe paper points to the direction from which the wind is coming. The crepe paper weather vanes may be used for creative dance experiences.

Let the children use the wind by flying a kite, by sailing a toy sailboat, or by playing with a pinwheel. See Figure 3–3 for direc-

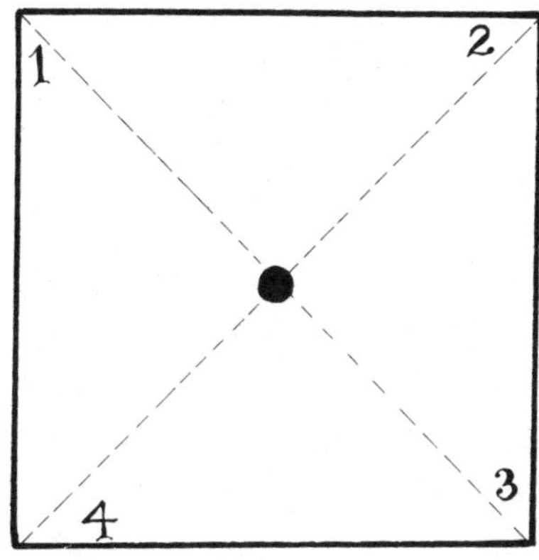

Cut a square of construction paper on the dotted line to the circle.

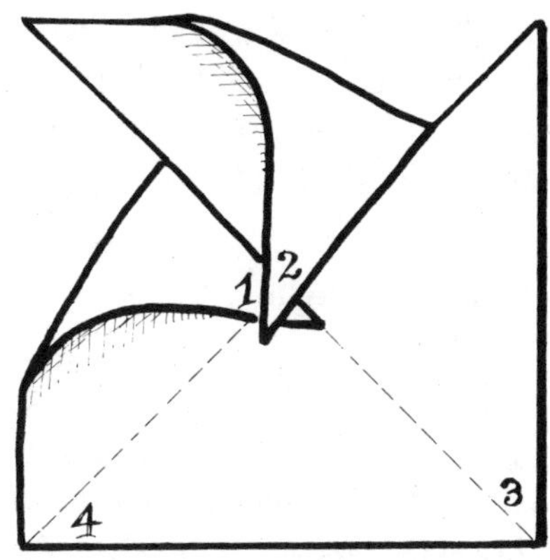

Fold corners 1, 2, 3, and 4 to the center of the circle in that order.

Figure 3-3
Pinwheel

tions on making a pinwheel. Cut a square of construction paper on the dotted line to the circle. Fold corners 1, 2, 3, and 4 to the center of the circle in that order. Run a large straight pin or hat pin through the center of the square after the four corners have been folded inward; then attach the point of the pin to a plastic drinking straw or a dowel. Discuss some ways people use wind to help them (to sail a boat, to pump water with a windmill, etc.)

Set up a weather station with a homemade weather vane, anemometer, rain gauge, and thermometer. The weather person of the day may report the wind direction and speed, the amount of rainfall, and the temperature.

Suggested Readings

Ets, Marie Hall. *Gilberto and the Wind.* New York: Viking, 1963.

Fontaine, La. *The North Wind and the Sun.* New York: Franklin Watts, Inc., 1964.

Kinney, Jean, and Kinney, Cle. *What Do the Clouds Do?* New York: Young Scott Books, 1967.

Knight, David C. *Let's Find Out About Weather.* New York: Franklin Watts, Inc., 1971. (For older children, but parts may be adapted for younger children.)

Martin, Bill, Jr. *Weather.* New York: Holt, Rinehart & Winston, 1967.

Miyumura, Layue. *I See the Winds.* New York: Thomas Y. Crowell, 1966.

Rosenfeld, Sam. *Ask Me A Question About the Weather.* New York: Harvey House, Inc., 1966. (For older children, but parts may be adapted for younger children.)

Rothman, Joel. *At Last, To the Ocean.* New York: Crowell-Collier Press, 1971.

Tresselt, Alvin. *Follow the Wind.* New York: Lothrop, Lee & Shepard, 1950.

Zolotow, Charlotte. *The Storm Book.* New York: Harper & Row, Pub., 1952.

EVAPORATION AND THE HYDROLOGICAL CYCLE

Information for Adults

The hydrological, or water cycle, consists of three parts: evaporation, condensation, and precipitation. Even though most of the time it may not be seen, water is always present in the atmosphere in one of three forms: gaseous, as invisible water vapor; liquid, as clouds, fog, or droplets of rain; and solid, as snow or ice. Water in the air comes from water surfaces, such as oceans and lakes, and from soil and plants by evaporation, a process by which the liquid form of water is changed by heat into a vapor (thus seeming to disappear into the air). Heat and wind accelerate the process of evaporation; the more of each there is, the faster water will evaporate.

Warmer air is able to hold more water than cooler air. For instance, a ten-foot cube or room can hold about five ounces of water at 40° F but can hold approximately twenty-two ounces of water at 80° F. When air becomes saturated with moisture—that is, when it contains all the water it can hold at its present temperature—it must give up some water. It gives up water through the process of condensation. Condensation may occur only if the air is cooled, as when the warm, moist air meets a layer of cooler air at higher elevations or when warm, moist air mixes with a mass of cooler air.

In addition to the need for cool air in order for condensation to occur, there is also a need for something for the water vapor to condense onto. In the air, water vapor may condense onto hygroscopic nuclei—small particles in the atmosphere that may be dust particles from the soil or the wearing away of rocks, carbon particles from factories and fires, or particles of salt from evaporated sea spray. Water vapor that condenses on hygroscopic nuclei may fall through the atmosphere as precipitation, in the forms of rain, snow, sleet, or hail.

Rain falls as water vapor and condenses to a liquid form, and small drops fuse to form drops of rain. Snow is formed when water vapor condenses into a solid form without first becoming a liquid. Sleet forms as rain falls through a layer of freezing air to become drops of ice. Hail begins as a frozen raindrop that keeps collecting layers of water. Additional layers are added until there is enough weight to the hailstone to allow it to fall to the earth. Dew and frost are forms of water vapor that condense directly onto cool objects on the ground and do not fall through the atmosphere. Dew is water vapor that has condensed at a temperature above 32° F, and frost is water vapor that condenses or sublimates (a more accurate term to indicate the change from a gaseous to a solid state) into a solid form at a temperature below 32° F.

Experience: Part I

PURPOSE: To discover that water disappears into the air.

MATERIALS: A chalkboard or dark-colored wall; a sponge filled with water; one small plastic bowl.

Let a child make a wet spot on the chalkboard or wall with the sponge, and when it disappears, question the children about what they have observed and about where they think the water has gone. They may suggest that the water has gone into the chalkboard or wall.

If it has not been previously introduced or if the children are unfamiliar with these basic facts about air, present the experience on air, "AIR: THE LUNGS—Experience: Part I." Ask whether there is a large amount of air in the entire room and whether there is only a little air in the small bowl. Have the child make two wet spots on the board or wall; then suggest that he or she leave all the air in the entire room around one spot but take most of the air away from the other spot by covering the spot with the bowl. Observe the uncovered spot disappear. Remove the bowl and discuss why this spot remained and the other disappeared. Explain that it is called evaporation when water disappears into the air.

Suggested Questions

1. After the one wet spot disappears, ask, "Where did the water go?"
2. After making two wet spots and covering one, ask, "Which spot has a lot of air around it and which has hardly any air?"
3. What happened to the spot with a lot of air around it? To the spot with most of the air taken away?
4. If the water went into the chalkboard or wall, would the spot without air have disappeared, too?
5. What was there all around the spot that disappeared? What did the water go into?

Experience: Part II

PURPOSE: To observe that heat makes water evaporate or go into the air faster.

MATERIALS: A source of strong light, such as a desk lamp; a sponge filled with water; a chalkboard or dark-colored wall.

Turn on the lamp and let the children feel that it is warm near the light source. Have one child make two wet spots with the sponge on the chalkboard or wall, and then hold the lamp near one spot in order to see which spot evaporates faster.

Suggested Questions

1. Which spot went into the air or evaporated faster?
2. What was different about the two spots?
3. Does heat make water go into the air or evaporate faster?

Experience: Part III

PURPOSE: To observe that wind makes water evaporate or go into the air faster.

MATERIALS: A piece of cardboard for fanning; a chalkboard or dark-colored wall; a sponge filled with water.

Let a child make two wet spots on the chalkboard or wall. Ask the child to make some wind by fanning or blowing on one of the spots to see which one evaporates faster.

Suggested Questions

1. Which spot went into the air or evaporated faster?
2. What was different about the two spots?
3. Does wind make water go into the air or evaporate faster?
4. What happens to the water in your washcloth when it sits overnight? Where does the water go?

Follow-Up Activities

Discuss where, in nature, one could find water that could evaporate into the air.

Let the children "paint" with water in the sunlight in order to see how quickly it evaporates. Let them test to see if it evaporates as quickly when they paint in the shade.

Let each child dip a finger into water in order to feel it evaporate. Have them dip two fingers into water; then ask how they could make one dry faster than the other. Does wind make water evaporate? How could you get some wind to dry your finger? Does sun make things warm? Could the sun help to make the water evaporate?

Suggested Reading

Holl, Adelaide. *The Rain Puddle.* New York: Lothrop, Lee & Shepard, 1965.

RAIN AND RAINBOWS

Information for Adults

As water evaporates, it goes into the air, where it is cooled and condenses on particles in the air to form a cloud. The tiny droplets of water in a cloud fuse to form larger droplets. When these droplets become large enough that all the water does not evaporate from them as they fall through the air, the droplets reach the ground in the form of precipitation known as rain. For more information, see "EVAPORATION AND THE HYDROLOGICAL CYCLE."

A rainbow is formed when light (usually sunlight) shines through water (usually raindrops). As the light passes through the water, it is both refracted (bent) and reflected (cast back). The light re-emerges from the water at an angle and splits, as with a prism, into the various colors of the spectrum. From the earth, this spectrum is seen as an arc with a band of violet on the inside, then blue, green, yellow, orange, and red on the outside. For more details, see "Information for Adults" in the unit on "LIGHT AND SHADOWS."

Experience: Part I

PURPOSE: To observe, in part, the process by which rain is formed.

MATERIALS: A source of boiling water; a pan filled with ice.

If the children are not familiar with the term *evaporation,* you may present the unit "EVAPORATION AND THE HYDROLOGICAL CYCLE." Boil the water in order to observe it evaporating into the air. Have the children feel the air near the pan filled with ice in order to confirm that this air is cold or cool. Hold the pan a few inches above the boiling water in order to see the "cloud" that forms as the water vapor from the boiling water meets the cool air near the pan. As you continue to hold the pan above the boiling water, the children will be able to see the water droplets condensing on the ice-filled pan, becoming larger, and finally falling downward. The closer the ice-filled pan is held to the source of boiling water, the faster the "raindrops" will fall. The adult may wish to help clarify what the children have seen by discussing how water evaporates in nature—where the cold air would come from, as there is no pan of ice in the sky, and what there would be for clouds and rain to condense onto.

Suggested Questions

1. What happens to the water when it boils? Where is it going? What is this called?
2. How does the air near the pan of ice feel?

3. What happens when the water that is coming from the pan of boiling water meets the cool air near the pan of ice?
4. As small water droplets join together to form larger droplets, what happens?
5. In nature, does water usually boil in order to evaporate? If not, what causes evaporation?
6. In nature, is there a large pan of ice in the sky? What might there be that could bring water out of the air?

Experience: Part II

PURPOSE: To discover that a rainbow is made when a light shines through water.

MATERIALS: A strong light source, such as a projector light; an aquarium filled with water; a piece of white paper; a water hose with a nozzle. (Note: If a projector light and an aquarium are not available, a clear glass filled with water and a strong beam of sunlight may be substituted.)

If using a projector light and an aquarium for this experience, darken the room slightly and shine the light through the water in the aquarium at an angle that creates a rainbow. The white paper should be placed in the spot where the rainbow is directed onto the table in order to make the colors appear brighter and clearer. If a glass of water and sunlight are used, place the white paper on the table or floor in the beam of light and hold the glass in the sunlight over the paper until the position that creates a rainbow on the paper is discovered.

Let the children give their suggestions about what made the rainbow; then let them experiment with removing the light source and with moving the water from the direct line of the beam of light. Ask them what is necessary in order to make a rainbow. Then ask them, "Where must the light be shining in relation to the water?"

On a sunny day, let the children spray a fine mist of water from the nozzle of the hose into the sunlight in order to create a rainbow. Discuss the three factors required to create this rainbow: light, water, and light shining through the water. Discuss where the light and water come from in nature when a rainbow is seen in the sky.

Suggested Questions

1. What made the rainbow?
2. What happens when we take the light away? When we take the water away?
3. Where, in relation to the water, must the light be shining in order to make a rainbow (over, under, around, through)?
4. In nature, from where does the light and water come to make a rainbow?

Experience: Part III

PURPOSE: To learn that the colors found in a rainbow always appear in the same order: violet, blue, green, yellow, orange, and red.

MATERIALS: The materials listed in Part II; strips or arcs of paper in the following colors: violet, blue, green, yellow, orange, and red; a large piece of white paper.

Create a rainbow indoors by one or both of the methods suggested in Part II and have the children arrange the strips or arcs of colored paper in the same order as the colors of the rainbow. Glue the strips or arcs of paper onto the piece of white paper.

Create a rainbow outdoors in order to let the children observe the order in which the colors are arranged. Compare the arrangement of colors to that of the colors glued to the paper.

Suggested Questions

1. What colors do you see in this rainbow?
2. In what order are the colors? Which color comes first, second, and so on?
3. Are the colors in the rainbow that we made inside the same as the colors of the rainbow we made outside? Is the order of the colors the same?

Follow-Up Activities

Make a rain gauge to measure rainfall. Directions for this activity may be found in "Experience: Part V" of "WEATHER MEASUREMENT."

Make rainbow cookies from the following recipe:

1 cup soft butter or margarine
1 cup brown sugar
2 eggs
1 teaspoon vanilla
3 cups unbleached or white flour
1/2 teaspoon baking soda
1/2 teaspoon salt
food coloring; red, yellow, and blue (mix these in the proper combinations to get orange, green, and violet)

Cream together the butter or margarine, sugar, eggs, and vanilla.

Mix together the flour, soda, and salt. Combine all ingredients and mix well.

Divide dough into six equal parts and color each part with food coloring to represent the six colors of a rainbow.

Pat each piece of dough into a strip about 4 inches long and 1/2 inch thick.

Stack all strips together and press gently to bond. Wrap in wax paper and refrigerate overnight.

Slice the layered dough into 1/8 inch slices, curve into an arc shape, and place on ungreased cookie sheet.

Bake at 350°F for about seven minutes. Makes 2 1/2 dozen. The idea for rainbow cookies is from the book, *I Can Make a Rainbow*, by Majorie Frank, published by Incentive Publications, Nashville, Tennessee, 1976.

As a listening activity for this unit, have the children listen to the record, *Clouds and Rainbows*, by Ann Esse Morrow, Educational Activities, Inc., Freeport, Long Island, New York, 1970.

In order to help children understand what thunder is, blow as much air as possible into two paper bags, holding the air inside the bags. Bump the two bags together with enough force to cause them to burst. This will illustrate the sound of thunder without lightning. In nature, a spark of electricity or lightning streaks through the sky, heating the air and causing it to expand rapidly. This expansion creates a shock wave, and the sound this wave makes is thunder. An explanation of the creation of lightning may be found in the unit titled "STATIC ELECTRICITY AND FRICTION—Experience: Part III."

Suggested Readings

Bartlett, Margaret Farrington. *Where the Brook Begins.* New York: Thomas Y. Crowell, 1961.

Black, Irma. *Busy Water.* New York: Holiday House, 1958.

Branley, Franklin. *Rain and Hail.* New York: Thomas Y. Crowell, 1963.

Edwards, Dorothy, and Williams, Jenny. *A Wet Monday.* New York: Morrow, 1976.

Freeman, Don. *A Rainbow of My Own.* New York: Viking, 1966.

McClosky, Robert. *Time of Wonder.* New York: Viking, 1957.

Ruskin, Karla. *James and the Rain.* New York: Harper & Row, Pub., 1957.

Tresselt, Alvin. *Rain Drop Splash.* New York: Lothrop, Lee & Shepard, 1964.

SEEDS

Information for Adults

All seeds come from reproductive parts in the center of flowers: the pistil—the part bearing the ovary—and the stamen—a stalk or column with a knobbed head where pollen is deposited—(see Figure 3-4). Most flowers have both parts in one flower, although in some cases the stamen and pistil grow in separate flowers. In the process of pollination (which is aided by wind and insects), pollen from the stamen, the male part of the flower, is deposited on the pistil, the female part of the flower.

The ovary, which contains the ovules or beginnings of the seeds, can take on many functions. It can be merely a pod to hold developing seed, or it can develop into a hard protective covering, as in the case of a cherry, peach, or plum seed. The ovary, in some cases, develops into a thickened covering with special adaptations for scattering seeds (barbs or spines) as on beggar's lice, sand spurs, or cockleburs; or wings may develop, as in the case of a maple, elm, or ash seed. The ovary can develop devices that propel seeds into the air as the pod bursts open, as in the case of wild geraniums, violets, or touch-me-nots. Seeds of milkweed, dandelions,

Figure 3-4

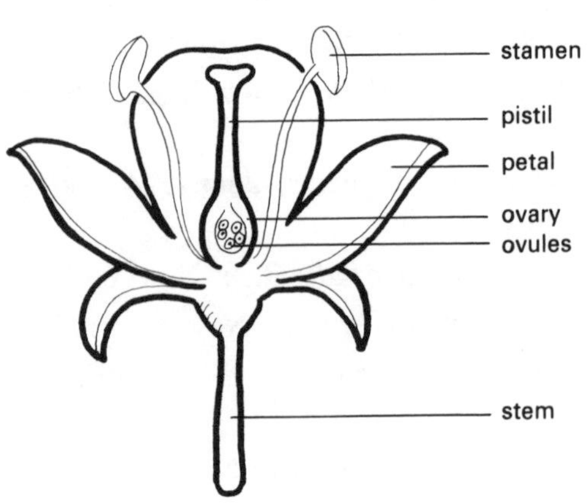

and sycamores develop silky hairlike structures that act as parachutes to suspend them in the air for a longer time so that the wind may catch them to carry them greater distances. Helpful illustrations dealing with how seeds travel may be obtained in a flannelboard kit from Instructo Products, Matthew and Cedar Hollow Road, Paoli, Pennsylvania, 19301. The kit is entitled "Seeds and Their Travels," Number 263.

In many cases, the ovary turns into a fleshy fruit, such as that of a tomato, cucumber, or squash. These examples illustrate that not all foods generally considered vegetables are considered so by scientists. In order to avoid confusion for the children, it is probably better to continue referring to fruits and vegetables as they are commonly classified. Some vegetables that do not contain seeds inside the fleshy or leafy part are carrots, radishes, cabbages, and lettuce. If these plants remain in the garden beyond the proper time for eating, they will grow a stalk upon which flowers, and then seeds, develop.

Within the seed coat, called the testa, is the embryonic plant surrounded by the endosperm or stored food for the plant. When a seed absorbs moisture, a chemical reaction occurs that triggers the growth of the embryo, helping it to break through the seed coat. Roots follow the pull of gravity and grow out of the seed downward. The embryonic plant resists the pull of gravity and grows upward. The endosperm nourishes the plant until it can survive on its own.

Experience: Part I

PURPOSE: To observe the parts of a flower that produce seeds.

MATERIALS: A collection of flowers that have been cut through the center with a sharp knife, as shown in Figure 3-4; a diagram of the inside of a flower, also found in Figure 3-4.

Show the diagram of the inside of a flower and point out the labeled parts. Display the flowers that have been cut open and help the children to find the pistil and its parts, the ovary and the ovules, and the stamen. Discuss pollination in terms of wind and insects moving the stamen to help deposit pollen on the pistil and the pollen traveling through the shaft of the pistil to the ovary where the ovules are pollinated in order to produce seeds.

Suggested Questions

1. Does the flower have a pistil? Where?
2. Where are the ovary and the ovules?
3. Where is the stamen?
4. What could cause the pollen from the stamen to get into the pistil where it then travels to the ovules?
5. Were you able to find a stamen and a pistil in every flower?
6. How do you think a flower that does not have a stamen and a pistil could become pollinated?

Experience: Part II

PURPOSE: To discover where seeds may be found.

MATERIALS: Fruits and vegetables that contain seeds; if possible, some vegetables that do not contain seeds and the stalk on which seeds of these plants grow; a variety of seeds in packets purchased from a garden or variety store; a knife.

Show the children the fruits and vegetables and ask where they came from and how they were grown. Discuss how the plant or tree that grew these fruits or vegetables started or ask where the seed that grew the plant or tree came from. Help the children to cut open the fruits and vegetables in order to discover the seeds inside. You may wish to explain that the fleshy part is the ovary of the grown or thickened plant. If they are available, show some vegetables without seeds as well as the stalk upon which the seed of the vegetable grows.

Use the seeds from the packets and the pictures of the plants on the packets to play a matching game. Let the children try to match each seed with the pictures of the plant that will grow from that seed.

Let the children classify the variety of seeds according to their own criteria. Discuss some other ways that the seeds may be grouped, such as by color or size or by whether they came from a fruit or vegetable or from a plant or a tree. Suggest that the children now try grouping their seeds in a different way.

Suggested Questions

1. Where did the plants that grew these fruits and vegetables come from?
2. From what did the plants grow?
3. Do all the fruits and vegetables have seeds inside?
4. In what ways are all these seeds different? Do any of the seeds from the fruits and vegetables look exactly alike?
5. Which of these pictures do you think shows the plant that this seed will grow into?

Experience: Part III

PURPOSE: To learn that seeds may be carried from place to place by people and animals.

MATERIALS: Some seeds with barbs or spines; pictures of animals that help seeds to travel (furry animals, such as dogs, rabbits, or foxes; squirrels; seed- and berry-eating birds; water birds; and earthworms).

Ask the children to suggest some ways that seeds are moved

about so that they grow in different places. Discuss that people may plant seeds on purpose or may accidentally scatter them as they pick flowers or throw the cores of fruits and vegetables on the ground after they've finished eating them.

Put one of the barbed or spiny seeds on a child's clothing and see if the children are now able to suggest another way that people help seeds to travel. Let them suggest some animals (usually furry) that could move seeds in this way.

See if the children are able to suggest some other ways that animals can help seeds to travel. Show pictures of animals that help seeds travel in order to help stimulate their thinking. For instance, a squirrel buries seeds as it stores food for the winter; a bird carries berries and seeds in its mouth and sometimes drops them on the ground; a bird also scatters seeds as it carries grass when building a nest; water birds, people, and some other animals carry seeds that stick to wet or muddy feet; and, as they "till" the soil, earthworms sometimes bury seeds that other animals have dropped.

Suggested Questions

1. Why do you think seeds may be found in so many places?
2. How do people help seeds to travel?
3. What are some ways in which animals help seeds to travel?
4. Could trucks or cars help to move the sticky types of seeds? How?
5. If you could design a twentieth-century flower, what would it look like? What devices would it have to help seeds to travel?

Experience: Part IV

PURPOSE: To discover that wind can help seeds to travel. To discover that water can help seeds to travel.

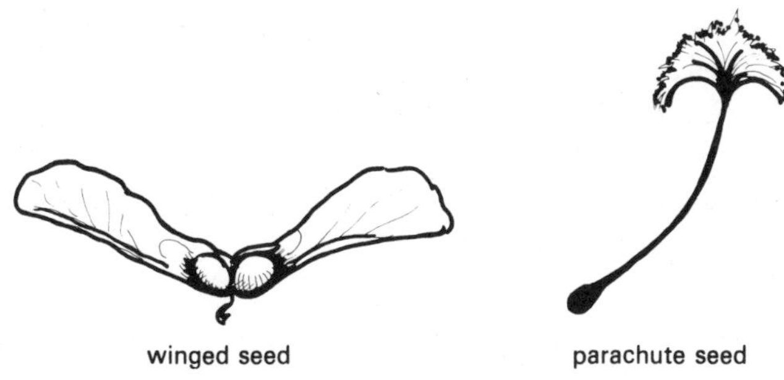

Figure 3-5

MATERIALS: Some small or lightweight seeds; a winged seed or a parachute seed (see Figure 3-5); a piece of paper or cardboard that may be used for fanning; a glass or squeeze bottle of water; a tray or plate.

Place some of the smaller or lightweight seeds on a tray or plate and have the children "make the wind blow" by fanning the seeds and observing what happens. Demonstrate how the special adaptations of winged seeds or parachute seeds can help them to travel very far in the wind. Toss one of these seeds into the air and, at the same time, toss up a regular seed: See which seed stays in the air longer to let the wind catch it and move it along. The children enjoy blowing winged or parachute seeds around the room.

Let the children pour some water over the small or lightweight seeds that are on the table in order to see that this causes the seeds to move or travel. Discuss where, in nature, the water would come from that helps seeds to travel.

Suggested Questions

1. When you made the wind blow in the room, what happened to the seeds?
2. Which seeds stayed in the air longer to let the wind help them travel?

3. What happened to the seeds when you poured water over them?
4. From where, in nature, would the water come to help seeds travel?
5. Who or what "plants" the seeds that grow into flowers in the woods and along the roadside?

Experience: Part V

PURPOSE: To discover how seeds turn into plants.

To observe a plant sprouting from a seed.

MATERIALS: Large seeds that have been soaked overnight in water (one for each child, plus some extras—dried lima beans work very well); a jar with straight sides; three or four paper towels.

Give each child a bean seed. Have them remove the seed coat and open the seed very carefully in order to see the embryonic plant inside. In order to sprout some of the soaked seeds in a glass jar, line the jar with damp paper towels and add about one inch of water to the bottom of the jar. Place several of the soaked seeds between the jar and the paper towel (the beans should stick to the sides of the jar). In two to three days the seeds will sprout, revealing the small plant and roots (see Figure 3-6). If the children do not already know the parts of a plant, show them the roots, the stem, and the leaves.

Suggested Questions

1. What did you find inside the seed when you opened it?
2. Does this give you any ideas about how a plant grows from a seed?
3. What came out of the seeds that we "planted" in the jar? Did someone open the seed or did it open on its own?

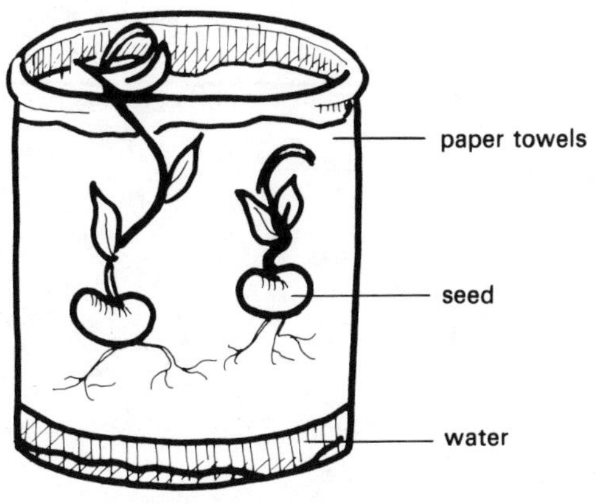

Figure 3-6

4. What do you think caused the seed to open and sprout? Do dry seeds sprout?

Experience: Part VI

PURPOSE: To observe small seeds sprouting.

To observe that, regardless of how a seed is planted, the roots will always grow downward and that the plant will always grow upward.

MATERIALS: Two small window panes (from a hardware store); a piece of blotter paper or construction paper cut to the size of a window pane; masking tape; small seeds (radish seeds work well); a glass jar with straight sides; paper towels; three soaked seeds (see Part V).

Help the children prepare for sprouting some very small seeds by placing the blotter paper or construction paper on top of one of the window panes and lining up several small seeds turned in various positions across the center of the paper. Place the other window pane over the seeds and seal three edges of the window panes with masking tape (see Figure 3-7). Pour

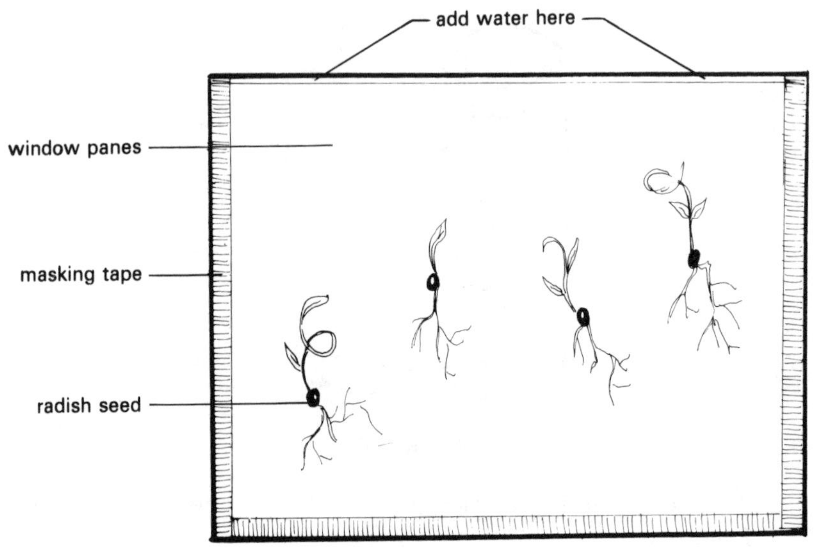

Figure 3-7

water into the unsealed edge. The seeds will sprout in several days if kept moist, but not wet.

Repeat the sprouting of the bean seeds in the jar with paper towels as described in Part V, but this time, place one bean upright, one upside down, and the other sideways in order to discover that the roots will always grow downward and that the plant will always grow upward, regardless of how the seed is turned.

Suggested Questions

1. What came out of the seeds when they were "planted" between the panes of glass?
2. What came out of the bean seeds in the jar?
3. Were all the seeds in the jar and between the panes of glass "planted" in the same way?

4. Which way did the roots of all the seeds grow? Which way did the plants grow?

Follow-Up Activities

For a snack, let the children eat some seeds (pumpkin, sun flower, nuts) and the fruits and vegetables in which they found seeds.

Provide the children with potting soil (this may be mixed with sand or gravel in order to avoid overwatering) and small paper cups or the bottom of styrofoam egg cartons so that they may plant as many different kinds of seeds as possible and watch them sprout. As they compare the variety of plants, they will begin to understand that a seed can only produce the same kind of plant from which it came (radish seeds can't produce carrots, etc.). Note: For this activity, try to provide seeds with a short germination period, such as lettuce, spinach, any member of the greens family, carrots, and radishes.

Listen to "How Do the Seeds of Plants Travel" on the record, *Now We Know, (Songs to Learn By)* by Tom Glazer, produced by Columbia Records.

Help the children to grow seeds in a mini-greenhouse that consists of two clear plastic disposable tumblers. Place wet cotton balls or some moist potting soil in the bottom of one tumbler. Put two or three rapidly germinating seeds on top of the cotton balls or just under the surface of the soil. Place the other tumbler upside down over the first tumbler (the tumblers should be placed rim to rim). Tape the tumblers together along the rims with clear cellophane tape or with masking tape. Put the greenhouse in light, but not in direct sunlight. The seeds will sprout in two to ten days, depending on germination time, and they should not need watering.

Grass seed may be sprouted in half an eggshell containing moist potting soil. Draw a picture of a face on the outside of the eggshell. As the grass grows, it becomes the "eggshell person's" hair, which may be cut as it grows longer. Individual cups to

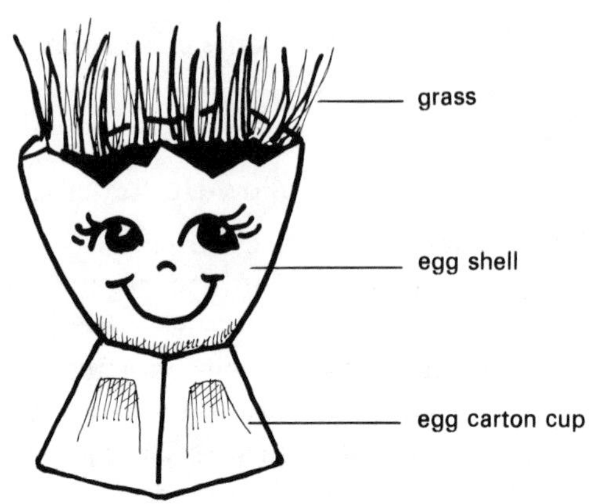

Figure 3-8

hold the eggshells may be cut from egg cartons. Turn the cups upside down. Cut a circle from the top and place the egg in the circle cut-out (see Figure 3-8).

Bean sprouts and grass seed may be grown by sprinkling the seed on a wet sponge that is sitting in a saucer. Place the saucer and sponge in light, but not in direct sunlight.

Suggested Readings

Black, Irma Simonton. *Busy Seeds.* New York: Holiday House, 1970.

Carle, Eric. *The Tiny Seed.* New York: Thomas Y. Crowell, 1970.

Collier, Ethel. *Who Goes There In My Garden.* New York: Young Scott Books, 1969.

Downer, Mary Louise. *The Flower.* Reading, Mass.: Addison-Wesley, 1955.

Harris, Petie. *The Seed the Squirrel Dropped.* Englewood Cliffs, N.J.: Prentice-Hall, 1976.

Jordan, Helen J. *How a Seed Grows.* New York: Thomas Y. Crowell, 1960.

——. *Seeds By Wind and Water.* New York: Thomas Y. Crowell, 1962.

Krauss, Ruth. *The Carrot Seed.* New York: Scholastic Book Services, 1971.

Selsam, Millicent E. *Seeds and More Seeds.* New York: Harper & Row, Pub., 1958.

PLANTS

Information for Adults

Although most plants grow from seed, some reproduce by modifying a portion of the plant, usually the stem or root, in order to produce a new plant. "Cutting" is the name given to a piece that is taken from the "mother" or original plant in order to produce a new plant. A plant should be well watered before a cutting is removed, and the cutting should be taken from relatively new growth, never from a section where the stem is becoming woody (turning hard or brown). A cutting may be from two to six inches in length and should be made with a sharp knife or razor blade. Some plants that reproduce well from cuttings are coleus, Swedish ivy, and wandering Jew. Cuttings from these plants will grow roots when placed in a glass of water, so that the children may observe the root development.

In some plants, portions of the stem modify by enlarging in order to reproduce. Enlarged underground stems, such as tubers, may be planted in order to produce a new plant. Some examples of tubers are Irish potatoes, sweet potatoes, and tuberous begonias. Bulbs, which are usually thought of as enlarged roots, are actually plant buds containing an entire plant. These bulbils are modifications for reproductive purposes. When bulbs are grown indoors to bloom early, they are said to be forced. Some easily grown bulbs include paper white narcissus, small-cup daffodils, hyacinths, and onions.

In order to grow, most plants need light, water, soil, and warmth. Light and water are necessary for photosynthesis, the process by which a plant uses light and chlorophyll to combine water and carbon dioxide chemically in order to produce its food, glucose. The soil supports the root system through which water is conducted and by which the plant is anchored firmly in a pot or in the ground; the soil generally contains the nutrients and minerals necessary for plant growth. Although some plants require cool temperatures for certain stages of growth, plants generally enter a dormant period during cold weather. The plants the children will work with require a certain degree of warmth for growth.

Enormous amounts of water may be pumped through plant tissues. Most of the water is released by transpiration (the use of water during photosynthesis), and a small percentage is used for plant nutrition. Water is absorbed through the fine tips of roots or rootlets (these usually have hairlike structures called root hairs) and moves upward in the plant.

Although there are many possible contributing factors to the upward movement of water in the plant, scientists generally feel that two factors—capillary action and cohesion—are mainly responsible. Capillary action refers to the movement of water in plant tissue by molecular attraction. Molecules of water adhere to the sides of a capillary or water-conducting vessel of the plant, pulling other molecules with them. This pulling action is created by cohesion, the attraction of water molecules for each other. In a long, slender column, such as the capillary of a plant, the cohesiveness of water has a much greater force than the downward pull of gravity.

For children, the absorption of water by a plant may be compared loosely to sucking a liquid through a straw. The suction begins at the upper portion of the straw in order to pull the liquid upward. Capillary action begins at the upper portion of the water column and pulls the water upward.

Experience: Part I

PURPOSE: To discover that new plants may be reproduced from cuttings.

MATERIALS: Plants that reproduce easily from cuttings, such as those suggested in "Information for Adults"; a sharp knife or razor blade (for adults' use only); a clear glass or jar filled with water (one for each cutting or type of cutting, if desired).

Remove one or more cuttings from each plant as the children watch. Let them place the cuttings in a jar or glass of water with only the bottom two or three inches of the stem submerged. Leaves should be removed from the submerged portion of the stem. In a few days (two to five for most of the plants recommended), roots will develop on the submerged portion of the stem. When an adequate root system has developed, the new plant may be planted in soil. Let the children compare the new or "baby" plant to the original or "mother" plant and discuss how the new plant was produced.

Suggested Questions

1. What is forming on the stem of the cutting that is in the water?
2. In what ways are the "baby" plant and the "mother" plant alike? Do the leaves look the same? Do both have roots? In what way or ways are the plants different?
3. What did we do in order to get a new plant from the old plant?

Experience: Part II

PURPOSE: To discover that new plants may be produced from tubers.

MATERIALS: A sweet potato or Irish potato; a jar or glass of water; toothpicks.

Discuss with the children that the sweet potato or Irish

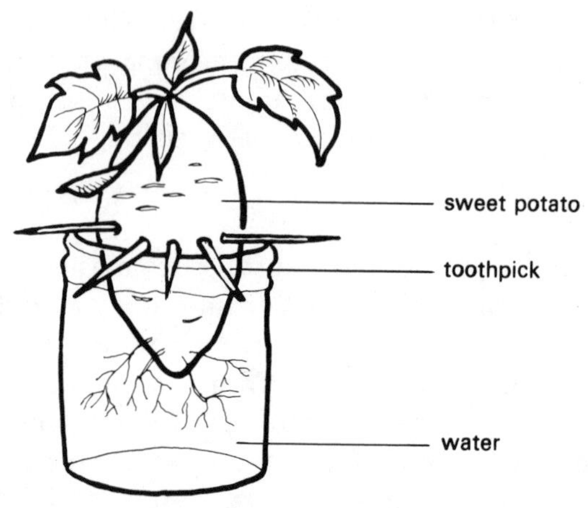

Figure 3-9

potato is an enlarged portion of a root and is called a tuber. Place toothpicks around the center of the length of a potato and submerge the bottom half by resting the toothpicks on the rim of the glass or jar (see Figure 3-9). It will take two or more weeks for hair roots (smaller or finer roots) and a new plant to begin growing from the sweet potato or Irish potato.

Suggested Questions

1. What do you see coming out of the part of the tuber that is under the water?
2. What is coming out of the part of the tuber that is not in the water?
3. How did we get a new plant from this tuber? What did we do first? Then what did we do?

Experience: Part III

PURPOSE: To discover that new plants may be reproduced from bulbs.

MATERIALS: Several bulbs that force easily, such as those suggested in "Information for Adults" (Do not choose bulbs that have already begun to send out shoots); a shallow bowl, such as a soup bowl; small pebbles (enough to fill the bowl 3/4 full); water (enough to just cover the pebbles in the bowl).

If bulbs have not been preconditioned for forcing, they must be placed in a refrigerator for four to six weeks, as they need to be exposed to a period of cold. After the bulbs have been conditioned, let the children place them on top of the pebbles that have been put in the shallow bowl. The bulbs must be placed right side up (the slightly indented, flat, or bald spot is the bottom; the pointed end is the top). The water level should be kept just barely covering the pebbles. Too much water touching the bulb will cause it to rot.

Suggested Questions

1. Why do the bulbs need to stay in the refrigerator for a while? In nature, are the bulbs exposed to cold?
2. What came out of the bulb first? What part of the bulb produced the roots? What part of the bulb produced the green stem and leaves?
3. How did we get a new plant from this bulb? What did we do first? Then what?

Experience: Part IV

PURPOSE: To discover what plants need in order to grow.

MATERIALS: Eight small plants—these may be purchased in flats from garden shops: Marigolds and salvia work well for this activity; a small pot or styrofoam cup with drainage holes for each plant; potting soil; a cardboard box.

Ask the children to give their ideas of what plants need in

order to grow. Suggest that they test out some of these ideas by taking the water, light, soil, and warmth away from a plant in order to see if it lives. Help the children plant seven of the eight small plants in pots or styrofoam cups that contain potting soil. The eighth plant should be placed in a pot or cup without soil.

In order to test whether or not a plant needs sun or light to grow, place two of the plants in a sunny window. Cover one plant with a cardboard box in order to remove the light. Both plants should be kept watered. Results will be evident in one or two weeks.

In order to test whether a plant needs water, place two plants in a location that receives sufficient light. Keep one plant watered, but not overwatered. Do not water the other plant. Results will be evident within a week.

In order to test whether a plant needs soil, place two plants, one with soil and the other without soil, in a location that receives sufficient light. Keep both plants watered. In less than one week, results will be evident.

In order to test whether a plant needs warmth, place one plant in a location that receives sufficient light and warmth and another plant in a refrigerator or freezer. Keep both plants watered. If one plant is kept in the freezer, results will be evident in one day. If the plant is kept in the refrigerator, results will be evident in about one week. Note: Because the plant in the refrigerator or freezer lacks both light and warmth, this is not a controlled experiment. The children will probably not realize this, as the light in the refrigerator or freezer will be on as they open the door. In order to achieve a controlled experiment, this experience may be presented in the very early spring where, in most locations, the weather is still quite cool. One of the plants may be placed outside in the cold where it also receives sunlight.

Suggested Questions

1. What happened to the plant that did not receive light? Does a plant need light?

2. What happened to the plant that did not have water? Does a plant need water?
3. What happened to the plant that did not have soil? Does a plant need soil?
4. What happened to the plant that did not have warmth? Does a plant need warmth?
5. From where, in nature, does a plant get light? Water? Soil? Warmth?

Experience: Part V

PURPOSE: To learn where a plant takes in water.

MATERIALS: Two plants in pots (bean plants or the plants used in Part IV); aluminum foil; a source of water.

Ask the children what part of the plant they think takes in the water. Suggest that they test in order to see whether the green part of the plant (the leaves and stem) or the roots under the soil takes in the water. Protect the roots of one of the plants from receiving water by molding aluminum foil around the stem of the plant, thus covering the top of the pot and completely covering the soil. Water only the upper part of this plant and only the soil (and thus the roots) of the other plant. After one week, compare the two plants.

Suggested Questions

1. What part of a plant do you think takes in the water?
2. What part of each of these plants are we watering—the roots or the stem and the leaves?
3. After one week, ask, "Which of these plants looks healthy? Which do you think has been getting water?"
4. Now what part of a plant do you think takes in water and keeps the plant growing?

Experience: Part VI

PURPOSE: To learn, in part, how a plant takes in water.

MATERIALS: A stalk of celery complete with leaves; a glass of water colored very dark with food coloring (blue and red work well); a knife; a clear plastic straw; a colored liquid that a child may drink.

Let the children add the coloring to the water so that they know that it is water in the glass but is different only because it has been colored. Explain that the celery is a plant that once had roots and show where the roots were. It is important to make a fresh cut in the celery by removing the bottom two inches of the stalk. Let the children place the celery in the glass of colored water with only the bottom of the stalk submerged. The colored water will move upward in the stalk until it has reached the tips of the leaves. The following day, the stalk of celery may be broken open with a crosswise break in order to reveal the veins through which the colored water has traveled.

The veins through which the colored water travels in the celery may be compared to small straws through which water moves. Let a child drink the colored liquid through the clear straw and compare this to the water moving upward in the veins of the plant and the child's sucking to the capillary action in a plant.

Suggested Questions

1. What happened to the colored water in the celery? What did the water travel through? Where did it go?
2. Did the colored water travel to all parts of the celery? Even to the leaves?
3. Where did the water enter the celery?
4. What happens to the liquid in the straw when you suck on the straw? Does the liquid move upward if you don't suck?

5. Does the liquid being sucked through the straw give you an idea about how a plant takes in water? (Let them explain.)

Follow-Up Activities

The plants that have been reproduced from cuttings make nice gifts. Let the children decorate styrofoam drinking cups with crayons (water color or marking pen decorations tend to get wet and smear) to use as pots for the plants. Be sure that there are drainage holes in the bottom of the cup. The "eye" or bud of an Irish potato or a tuberous begonia may be planted in a pot with potting soil. Keep them well watered, but not overwatered, and plants will emerge.

As the children observe plants growing around them, discuss whether these plants have light, water, soil, and warmth, and then ask from where, in nature, the plant gets each of these.

Flowers may be placed in colored water, as was the celery, in order to demonstrate that they also take in water from the bottom of the stem (the roots) upward. White flowers, such as carnations, mums, and daisies work well.

Carrots and turnips, both of which are enlarged roots but not tubers, may be used to make "hanging" plants. Cut about four

Figure 3-10

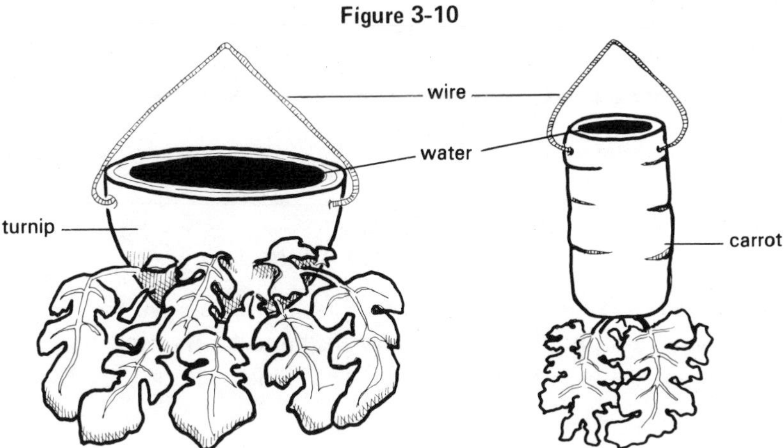

inches off the thick end of a large carrot or about one inch off the top of a turnip root (just enough to make one end of the turnip flat). Hollow out the four-inch piece of carrot or the large portion of the turnip root with a vegetable peeler. Hang either by forcing the ends of a length of wire through the root in order to form a handle, as on a bucket. Fill with water and keep filled. Greenery will grow from the sides of the root (see Figure 3-10).

Suggested Readings

Benton, William, and Benton, Elizabeth. *How Does My Garden Grow?* Racine, Wis.: Western Publishing Co., Inc., 1969.

Collier, Ethel. *Who Goes There in My Garden?* New York: Young Scott Books, 1963.

Darby, Gene. *What is a Plant?* Atlanta, Ga.: Benefic Press, 1960.

Dowden, Anne Ophelia. *Wild Green Things in the City—A Book of Weeds.* New York: Thomas Y. Crowell, 1972.

Downer, Mary Louise. *The Flower.* Reading, Mass.: Addison-Wesley, 1955.

Lubell, Winifred, and Lubell, Cecil. *Green is for Growing.* Chicago: Rand McNally, 1964.

Podendorf, Illa. *The True Book of Plant Experiments.* Chicago: Children's Press, 1960.

Webber, Irma. *Up Above and Down Below.* Reading, Mass.: Addison-Wesley, 1953.

THE ENVIRONMENT

Information for Adults

A person's environment includes his or her surroundings, both living and nonliving, and the influences affecting the way the person grows, develops, and lives. This unit on "THE EN-

VIRONMENT" and those that follow on "POLLUTION" and "ECOLOGY AND CONSERVATION" are designed to help the child to develop an appreciation for the environment (particularly for those surroundings that are not made by humans) and to understand that natural resources in the environment are finite and should not be abused.

One way to help to develop an appreciation for the natural environment is through nature study, particularly, as described in this unit, nature walks. Nature may be studied on a walk through the forest, an open field, a park, or even a vacant city lot. The adult may wish at times to give more definition to a nature walk by helping the children to discover things they might otherwise have overlooked. Experience: Part I gives some specific suggestions for nature walks. On all nature walks, it is wise to follow an adage often used in scouting: "Take nothing away but yourself. Leave nothing behind but your footprints."

Different plants and animals may require different environments, but all the plants and animals living in and sharing the same environment make up a biological community. There are many different types of biological communities, including forest communities, pond communities, desert communities, vacant lot communities, and more.

Sometimes, over several generations, plants and animals make modifications or adjustments to help them live in a changing environment. These adjustments or adaptations may be ones that help the plant or animal to fit more perfectly into its environment. Examples of such adaptations are aids for obtaining food or shelter, adaptations to aid reproduction and to offer protection, including defense mechanisms and camouflage.

Some anatomical adaptations that may be observed in animals include modification of claws, wings, tails, flippers, fins, snouts, feet, eyes, ears, teeth, and mouthparts or sensory adaptations. Humans often adapt to their environment by copying ideas of adaptations from other animals. Figures 3-11a–3-11h illustrate examples of ways that humans copy ideas from animals.

Figure 3-11a
Turtles have hard shells that serve as protection.

Figure 3-11b
People copy this idea with the football helmet, which acts as a hard shell in order to protect the skull.

Figure 3-11c
The spider constructs a web to capture food.

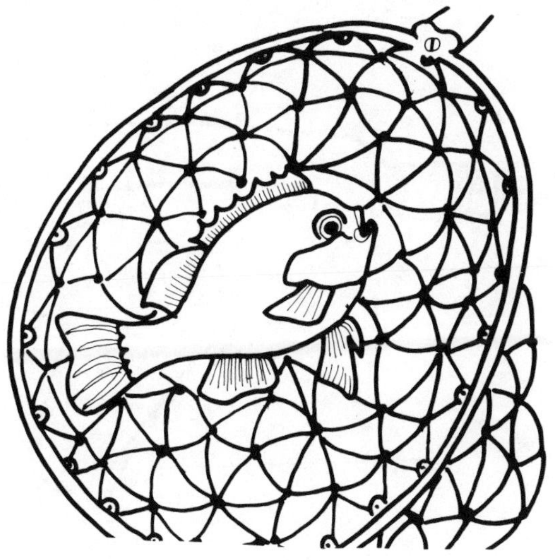

Figure 3-11d
People use a weblike fishing net that helps them to capture food.

Figure 3-11e
A long pointed bill enables woodpeckers to drill holes in trees and wood.

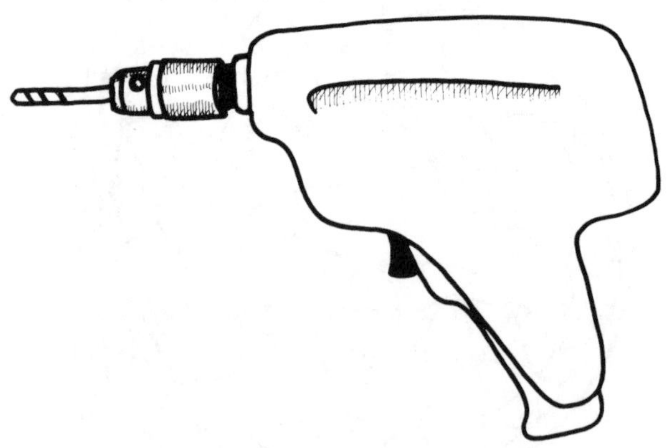

Figure 3-11f
People use a similarly shaped bit with an electric drill in order to make holes in wood.

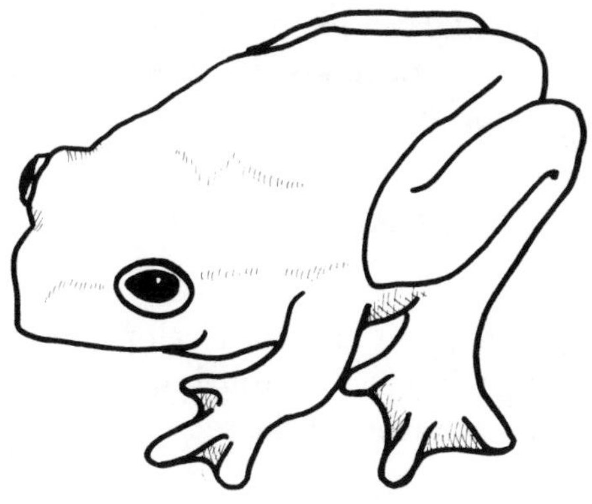

Figure 3-11g
The webbed foot of a frog may be used as an aid in swimming.

Figure 3-11h
People create their own webbed feet with flippers in order to aid them in swimming.

Scholastic Book Services offers teaching units and activity books on environmental awareness. The address for ordering is 904 Sylvan Avenue, Englewood Cliffs, New Jersey, 07632. The unit is called "Look Around You."

Experience: Part I

PURPOSE: To explore and to develop an appreciation for the natural or living environment through walks.

MATERIALS: Two feet; all the senses.

As an introduction to the study of the environment, take one or several nature walks, emphasizing different approaches for exploring nature. Following are specific suggestions:

Take a *listening walk.* Walk very quietly and listen for the sounds of nature. Stop at intervals and have the children close their eyes and listen. See if they are able to identify some of the sounds that they hear and to locate the direction from which the sound is coming. Let the children place an ear against a tree trunk or against the ground in order to listen for activity within.

Take a *touching walk.* Instead of just seeing nature, let the children feel it by exploring with the sense of touch. Find textures that are rough, smooth, sticky, soft, and hard. See if the children are able to find something warm, something cool, something wet, and something dry.

Take a *smelling walk.* Suggest that the children try to find good smells, bad smells, and sweet, spicy, or rotten smells. Let them smell leaves from many different trees and plants, and then ask them to crush a leaf from each between their hands in order to see if the odor changes in any way. Suggest that they smell the soil, a rotten log, under a pile of leaves, some wild flowers.

Take an *upside-down walk.* Instead of just looking at surfaces, look at the undersides of leaves, logs, rocks, and so on. See if the objects are different in color, texture, or appearance when turned upside down. Does anything live on the underside (insects or other animals, moss, mushrooms)? When the children have observed the underside of an object, have them return it as nearly as is possible to its original position. Note: The adult should caution the children about snakes or poisonous insects before beginning this activity.

Take a *color walk*. Try to find every color represented in nature. Look for yellow, violet, blue, white, red, or pink flowers; green leaves (see how many different greens may be found); brown straw, leaves, grass, and soil; black tree bark or soil; beige or grey mushrooms.

Suggested Questions

1. What new things did you discover on our listening walk, our touching walk, and so on?
2. What were your favorite things that you found on our walk?
3. Were your favorite things made by humans? Could they be made by humans?
4. Should we try to protect or save the natural things in our environment? Why or why not?

Experience: Part II

PURPOSE: To understand the meaning of the word *environment* and to compare different types of environments.

MATERIALS: Old magazines to use as a source of pictures illustrating different types of environments and different aspects of the children's environment (*Ranger Rick's Nature Magazine*, published by the National Wildlife Federation, and *World* magazine and *The National Geographic*, published by the National Geographic Society, are excellent sources of pictures).

Discuss with the children what is meant by the word *environment* (see "Information for Adults") and have them name some of the things in their environment, living and nonliving.

Show pictures of other types of environments in which children may live and let the children compare these environments to their own. The pictures should show some contrast, if possible. For example, a tropical and an arctic environment may be compared; a city and a country environment may be compared.

After a discussion of environments, let the children make a notebook entitled "Ten Beautiful Things In My Environment." The notebook may consist of some actual objects from the environment or of pictures drawn by the children or cut from magazines.

When the notebooks are complete, discuss with the children that many of the beautiful things in their notebooks are natural and not manmade. Discuss whether or not people could reproduce any of these natural items.

Suggested Questions

1. What are some things in your environment?
2. Which things in your environment are living and which are not? What makes something a living thing?
3. In what ways is the environment these children live in like yours? In what ways is it different?
4. How many of the beautiful things in your notebook are natural?
5. Could a person make any of these natural things?
6. From all the different environments we have discussed, take your favorite parts of each and create a new type of environment. What is this environment like?

Experience: Part III

PURPOSE: To understand what is meant by the term *community.*

To explore different types of plant and animal communities.

MATERIALS: Pictures illustrating different types of plant and animal communities. (*Ranger Rick's Nature Magazine*, published by the National Wildlife Federation, is an excellent source of pictures.)

Discuss what is meant by the term *community* (see "Information for Adults") and talk with the children about what makes up their community. Let them name some of the plants and animals, including people, that are living together and sharing their community.

Show pictures of different types of communities and discuss some of the plants and animals that live in these communities. Provide the children with backgrounds that you have drawn illustrating different types of communities—for example, a forest, a pond, a desert, or an ocean. Let the children fill in the pictures by drawing some of the animals that would share each community.

Let the children study a mini-community. Go into a wooded or natural area where plant and animal life is found. Draw a circle about four feet in diameter on the ground and look for evidence of plants and animals that live together in this area. Look for small animals, such as insects and spiders, and evidence of larger animals, such as tracks, chewed cones, nuts, feathers, or fur.

Suggested Questions

1. What plants and animals share your community?
2. Is this picture of a community anything like your community? In what ways is it like your community? In what ways is it different?
3. Here is a picture of the plants that share this community. What animals would share this community?
4. Could humans live in this community? Why or why not?

Experience: Part IV

PURPOSE: To learn the meaning of the word *adaptation.*

To observe some of the adaptations animals have made and to learn how adaptations may be useful to animals.

To discuss some ways that humans copy adaptations from animals.

MATERIALS: Pictures of animals that display some of the adaptations listed in "Information for Adults" (*Ranger Rick's Nature Magazine* and *World* are excellent sources for pictures); pictures from Figures 3-11a-h—colored, if desired, and cut into flash cards.

Show pictures illustrating the physical adaptations of some animals, and then discuss the ways the adaptations may be useful to the animal.

Ask the children whether or not humans ever copy ideas for adaptations from the animals. The pictures in Figures 3-11a-h may be used in this discussion to illustrate some ways that humans do copy the physical adaptations of animals.

After the discussion on adaptation, the cards from Figures 3-11a-h may be used as a lotto game. The children may match the adaptation of the animal with the pictures illustrating the way humans have copied the adaptation.

Suggested Questions

1. What adaptation does this animal have? What part of this animal is especially designed to help it live in its environment?
2. Does the adaptation help the animal? In what way?
3. Here is a picture of an animal with an adaptation. How have people copied this adaptation?
4. What does the adaptation do for the animal? How does it help? What does the copied idea do for the human? How does it help?

5. Think of an adaptation that you would like to copy from an animal. How could this be done?

Follow-Up Activities

Provide pictures of plants and animals that belong to various communities and let the children group them into the community to which each belongs.

Let the children use their imaginations by playing a "What could I do if I had . . ." game—for example: What could I do if I had the following:

1. A nose like an elephant.
2. A tail like a monkey.
3. Big ears like a rabbit.
4. Wings like a bird.
5. Fins and gills like a fish.
6. A neck like a giraffe.

Suggested Readings

General:

Carrick, Carol, and Carrick, Donald. *A Clearing in the Forest.* New York: Dial Press, 1970.

——. *Swamp Spring.* New York: Macmillan, 1969.

Tresselt, Alvin. *The Dead Tree.* New York: Parent's Magazine Press, 1972.

Nature Walks:

Behnke, Frances. *What We Find When We Look Under Rocks.* New York: McGraw-Hill, 1971.

Bolognese, Don. *All Upon a Stone.* New York: Thomas Y. Crowell, 1971.

Busch, Phyllis. *Exploring As You Walk in the City.* Philadelphia: Lippincott, 1972.

Fisher, Aileen. *Once We Went on a Picnic.* New York: Thomas Y. Crowell, 1975.

Howell, Ruth. *A Crack in the Pavement.* New York: Atheneum, 1970.

Hurd, Edith Thatcher. *This is the Forest.* New York: Coward, McCann & Geoghegan, 1969.

Sarton, Mary. *A Walk in the Woods.* New York: Harper & Row, Pub., 1976.

Showers, Paul. *The Listening Walk.* New York: Thomas Y. Crowell, 1961.

Animal Adaptations and Defenses:

Bowman, John S. *On Guard: Living Things Defend Themselves.* Garden City, N.Y.: Doubleday, 1969.

Frith, Michael. *Some of Us Walk, Some Fly, Some Swim.* New York: Random House, 1971.

Russell, Solveig Paulson. *What Good is a Tail?* Indianapolis: Bobbs-Merrill, 1962.

Selsam, Mellicent E. *Hidden Animals.* New York: Harper & Row, Pub., 1969.

POLLUTION

Information for Adults

In this unit, pollution refers to wastes contributed by humans that degrade the quality of the environment. In the unit, the children will begin to understand that only humans can stop pollution or clean up the environment. There are four types of pollution that may be studied with the children: air pollution, water pollution, solid waste pollution, and noise pollution.

Polluted air is air that has been made unclean by products (contributed by humans) that are unnatural or do not belong in this part of the environment in large concentrations. Most air pollution is caused by industries and the internal combustion engine (cars, trucks, buses). These put hydrocarbons, nitrogen oxides, and other harmful chemicals into the air. Sunlight acts on the hydrocarbons and nitrogen oxides in order to produce "smog." Smog can damage the lungs and heart and can irritate the eyes and nose. Air pollution can reduce the amount of sunlight received in an area and can change the temperature of the air. These factors may affect growing plants.

Polluted water is water made unclean by products (contributed by humans) that are unnatural or do not belong in this part of the environment. Most water pollution is caused by industries that dump harmful wastes, including poisons, into lakes and rivers. Another chief cause of water pollution is the use of detergents for household upkeep. Most detergents are high in phosphates, compounds that can act as fertilizers to speed up the growth of algae. Excess algae can cut off the oxygen supply to fish and other water animals. Polluted water is unsafe for drinking or for other human or animal use. It kills water animals and indirectly harms humans who eat water animals that have been affected by water pollution.

Solid waste pollution, usually called "litter," makes the environment unhealthy and unattractive. It consists of products contributed by humans to the environment. These are highly refined products not normally found in the natural environment and include plastics, cans, paper, glass, and other solid waste and garbage. Litter attracts rats and other disease-carrying animals.

Noise pollution is created by too many manmade sounds in the environment. It is made by noises from large machinery, riveters, drills, sirens, airplanes, car horns, even music that is played too loudly. Noise pollution can be injurious to hearing and may affect the nerves, making people irritable. It may even affect general health.

The David C. Cook Publishing Company, 850 North Grove Avenue, Elgin, Illinois, 60120, offers an excellent group of teaching pictures that may be used with this unit. The packet is entitled "Ecology: The Pollution Problem." Scholastic Book Services Inc., 904 Sylvan Avenue, Englewood Cliffs, New Jersey, 07632 offers Environmental Awareness units that include activity books and information on pollution. The unit is called "Look Around You."

Experience: Part I

PURPOSE: To become aware that air pollution exists and to understand, in part, what causes it.

To understand that only people can stop air pollution and can clean it up.

MATERIALS: Pictures illustrating air pollution and its causes (those suggested in "Information for Adults" are excellent); a piece of cardboard and a piece of white paper (both approximately 8″ × 10″); glue; vaseline.

Use the pictures to discuss what air pollution is and what causes it. Discuss with the children some ways in which air pollution is undesirable or harmful.

In order to illustrate that air pollution exists where the children live and attend school, set up a pollution index card (glue the white paper to the cardboard and cover the paper with a coat of vaseline). Place the card outside near a busy street or near where parents pick up their children after school. Leave it overnight. The next day, observe the particles that have collected on the vaseline-coated paper. Explain that these particles came from the air and that the cars, trucks, or buses probably contributed a great deal to the air pollution. Note: Dirt particles that collect on the paper need not be distinguished as different from pollution particles.

Discuss with the children some things that they could do to help stop air pollution. Some examples are as follows: (1) Do not ride in a car or bus when you can walk or ride a bicycle. (2) Do not build fires outdoors to burn leaves or trash. (3) Plant trees and other green plants that "freshen" the air by adding oxygen. (4) Save our forests (which add oxygen to the air) by putting out campfires and by helping to recycle paper. (5) When you become an adult, do not smoke; explain that smoking pollutes the air.

Suggested Questions

1. What is wrong with the air in this picture? What is causing this air to be dirty?

2. Who drives the cars, trucks, and buses and who runs the factories?
3. Would you like it if the air around you were polluted like this?
4. What is wrong with air pollution?
5. Did our pollution index card show evidence (or signs) of air pollution? What could have caused this pollution?
6. Who is polluting the air? Who can stop air pollution?
7. What can you do to help stop air pollution?

Experience: Part II

PURPOSE: To become aware that water pollution exists and to understand, in part, what causes it.

To understand that only people can stop water pollution and can clean it up.

MATERIALS: Pictures illustrating water pollution and its causes (those suggested in "Information for Adults" are excellent); two quart jars (with lids) half filled with water; detergent; fabric softener; bleach; grease (or any combination of liquids or powders commonly used in the home); three or four samples of detergents; one pint jar (with a lid) half filled with water for each detergent.

Use the pictures to illustrate what water pollution is and what causes it. Discuss with the children some ways in which water pollution is undesirable and harmful.

In order to illustrate to the children that water is being polluted by products from their homes, pour detergent, bleach, grease, and so on into one of the jars of water, place the lid on the jar, and shake. Compare the "polluted" water in this jar to the jar of clean water. Ask the children which kind of water they would rather drink, bathe in, or swim in.

In order to begin to help the children understand how they

may help reduce water pollution at home, make a comparison test of several different types of detergents. Place a small amount of each detergent in a jar of water. Screw the lid onto the jar and shake well. The detergent that produces the least suds is generally the least polluting.

Discuss with the children some things they could do to help stop water pollution. Some examples are as follows: (1) Do not waste the clean water that we do have. Turn off the water when you finish using it. Take quick showers or fill the tub with only two or three inches of water when bathing. (2) Never throw trash into a river, lake, pond, or the ocean. (3) Ask your parents to run the dishwasher or washing machine only when it is full, to use a low-sudsing or phosphate-free detergent, and to purchase white toilet tissue, since the dye in colored tissue adds to water pollution.

Suggested Questions

1. What is wrong with the water in this picture? What caused this water pollution? Who put the detergent, chemicals, or oil into this water?
2. Would you like the water that you drink, bathe in, or swim in to be polluted like this?
3. What is wrong with polluted water?
4. What kind of things did we add to the one jar of water? Do you use any of these things at your house?
5. What did the detergent and chemicals do to the water?
6. Who pollutes the water? Who can stop water pollution?
7. What can you do to stop water pollution?

Experience: Part III

PURPOSE: To become aware that solid waste pollution exists and to understand, in part, what causes it.

To understand that only people can stop solid waste pollution and can clean it up.

MATERIALS: Pictures illustrating solid waste pollution (those suggested in "Information for Adults" are excellent); contrasting pictures illustrating unpolluted parks, beaches, forests, and so on; small paper bags (one for each child); a large cardboard box or garbage can.

Use the pictures to discuss with the children what solid waste pollution or "litter" is and what causes it. Compare the pictures of polluted areas with those of unpolluted areas. Ask the children which of the areas look more beautiful and which seem healthier as places to live and play. Discuss some ways that litter is undesirable or harmful.

In order to illustrate that solid waste pollution exists where the children live and attend school, have them take the paper bags to gather up litter on the school grounds, in a public park, or in their neighborhoods. All the litter should be dumped into the cardboard box or garbage can in order to illustrate the large amount that has probably been gathered. This project may be continued for several days. The children enjoy giving names, such as "litter lover" or "garbage grabber," to the box or garbage can.

Discuss with the children some things that they can do to help stop solid waste pollution. Some examples are as follows: (1) Always throw litter into the nearest trash can. (2) Never throw trash out of a car or bus window and never leave trash in public areas. (3) Pick up litter that "litterbugs" have left. (4) Make posters telling others not to litter.

Suggested Questions

1. What is wrong with this picture? What caused this solid waste pollution? Who left the litter in this area?
2. Would you like to live in or near an area that looks like this? What is wrong with solid waste pollution?

3. Did you find much litter on the school grounds, in the park, and so on? Who left the litter there?
4. Who makes solid waste pollution? Who can clean up solid waste pollution?
5. What can you do to stop solid waste pollution?

Experience: Part IV

PURPOSE: To become aware that noise pollution exists and to understand, in part, what causes it.

To understand that only people can stop noise pollution.

MATERIALS: Pictures that illustrate some causes of noise pollution (those suggested in "Information for Adults" are excellent); a portable tape recorder; a tape or cassette.

Use the pictures to discuss with the children what noise pollution is and what causes it. Discuss some ways that noise pollution is undesirable or harmful.

You may use the tape recorder to record some of the sounds that cause noise pollution—car horns, bulldozers or other large machinery, riveters, drills, sirens, lawn mowers, motorcycles, airplanes, large trucks, and the like. Play the recording and let the children identify what is making the noise in each case. These noises may be compared to sounds in nature, such as rushing water, rain, birds singing, rustling leaves, and so on. (This may be done by recording the sounds of nature or by going to a quiet natural setting.) Ask the children whether any of the noises hurt their ears and ask which were the loudest noises and the softest noises. Discuss whether it would be more pleasant to live in a place that had much noise pollution or little noise pollution and ask which place would be easier to sleep in.

Discuss with the children some things that they can do to help stop noise pollution. Some examples are as follows: (1) Do not honk car horns or bicycle horns when unnecessary. (2) Do not play record players, television sets, radios, or musical instruments too loudly. (3) Do not shoot off fireworks and capguns.

Suggested Questions

1. What are some things in this picture that cause noise pollution? Who drives the truck; who plays the musical instruments?
2. What is making the noise pollution in this recording?
3. Which noises are more pleasant—those made by machines built by humans or those made by things of nature?
4. Would you prefer to live and play in an area that is polluted with noise or in one that has little noise pollution?
5. What can you do in order to reduce noise pollution?

Follow-Up Activities

Read "A Story About Pollution," found in Appendix II. After reading the story, let the children use their knowledge from this unit to tell how the people can clean up "Cleantown."

Let the children make "litterbug litterbags" to use for collecting litter. The litterbug face is made by gluing construction paper on the bag to make a face, as shown in Figure 3-12. (The adult draws the features on paper, and the children cut and paste them in the proper places.) When the bag is turned right side up and handles of string or heavy paper are attached, it becomes a litterbag.

In order to illustrate how much unnecessary solid waste pollution is created, have the children help to unpack groceries after the next shopping trip. Suggest that they put all the wrappings to be thrown away in one pile. Then determine how many of these wrappings were really necessary. How many were made of plastic (an extremely polluting material) when they could have been made of recycled paper? How could some of the "necessary" solid waste be reduced? Could grocery bags be reused? Could card-

litterbug

litterbag

Figure 3-12

board boxes for eggs, fruit, and vegetables be reused? Could bottles be reused?

Suggested Readings

Bartlett, Margaret Farrington. *The Clean Brook.* New York: Thomas Y. Crowell, 1960.

Milgram, Harry. *ABC of Ecology.* New York: Macmillan, 1972.

Parnall, Peter. *The Mountain.* Garden City, N.Y.: Doubleday, 1971.

Peet, Bill. *Farewell to Shady Glen.* Boston: Houghton Mifflin, 1966.

——. *The Wump World.* Boston: Houghton Mifflin, 1970.

Smaridge, Norah. *Litterbugs Come in Every Size.* Racine, Wis.: Western Publishing Co., Inc., 1972.

ECOLOGY AND CONSERVATION

Information for Adults

Ecology is the study of the relationship of all living things to each other and to their environment. An ecosystem includes the living things and the nonliving elements (soil, water, air) in a given area. A natural system of checks and balances exists within ecosystems that essentially maintains constant proportions of plants and animals in a given ecosystem and facilitates the flow of energy necessary in order to maintain life. Among these checks and balances are included natural enemies, climate, weather patterns, natural disasters, disease, and migration.

When humans tamper with any part of the ecosystem, this dynamic balance may be disrupted. An example of this type of disruption was the wide use of DDT, a pesticide that, although it killed most garden or farm pests, also killed predators that fed on these pests. The pests not killed by DDT had no natural enemies to destroy them and reproduced offspring showing immunities to DDT. Thus a destructive cycle was created in which more DDT and stronger forms of DDT had to be used until the poison was finally banned.

The energy of the sun supports all living things on the earth. Only plants are able to use the sun's energy directly, but the energy may be transferred from one organism to another. Thus, animals use the sun's energy indirectly as they eat plants or other animals that eat plants. Some of the energy is lost each time it is transferred to another organism. Humans use the sun's energy more efficiently by eating plants instead of the animals that feed on plants.

A food chain demonstrates the passage of food energy through a series of organisms. The energy for all food chains begins with the sun and is passed to green plants. The food energy of the plant is passed to a primary consumer (an animal that eats the plant), then to a secondary consumer (an animal that eats the primary consumer), and then possibly to other consumers. Some food chains have as many as five links, but most have only three or four.

Elements from the soil, water, and air are not as easily renewed as the sun's energy. There is a limited supply of elements, but, in nature, they are able to circulate from nonliving things to living things and back to nonliving things. This cycle may be demonstrated within a food chain. When any organism of the chain dies, it decomposes, and the elements are then recirculated or recycled in the soil, air, or water. These nonliving elements are responsible for continuing the chain by supporting new life.

Conservation is the wise use of natural resources, the products produced by nature that cannot be supplied by humans. These natural resources may be those necessary for life, such as clean air, clean water, and plants, or those resources that improve the quality of life, such as fuel oil and natural beauty. Some natural resources, such as oil, gas, and coal, are limited in supply and cannot be renewed on a rapid time scale. Other resources, such as trees, animals, and clean water, are more easily renewable, and an adequate supply may be maintained with careful use. Any study of conservation with children should stress the need for a lack of human interference with natural ecological processes or should explain that carefully controlled intervention may be used only when necessary.

Scholastic Book Services, Inc., 904 Sylvan Avenue, Englewood Cliffs, New Jersey, 07632 offers units and activity books on ecology and conservation. The unit is called "We Need Each Other."

Experience: Part I

PURPOSE: To demonstrate that once humans destroy the patterns or plan of nature, it is extremely difficult to repair the damage.

MATERIALS: A "frame" of construction paper for each child (cut all but the outer inch of paper away, as if cutting a mat for a picture).

Give each child a paper frame and go into a wooded or natural area. Have the children place their frames on the ground and carefully observe how objects within the frame are arranged in order to form a "picture." Tell them to remove the frame and destroy what made up the "picture" with their hands or feet. Now replace the frame and try to make the "picture" in the frame look exactly as it did before it was destroyed.

This activity may bring up a discussion of some other ways in which humans can destroy or interfere with nature to a point where the damage is irreparable.

Suggested Questions

1. What is in the "picture" inside your frame? What do you see?
2. Were you able to repair your picture after you destroyed it so that it looked exactly the same?
3. Do you think that humans are able to completely repair the damage once they have destroyed something in nature?
4. Can you think of some ways in which humans interfere with or damage nature?

Experience: Part II

PURPOSE: To discover what is meant by a food chain and how human interference can disrupt ecological systems.

MATERIALS: The food chain pictures found in Figures 3-13a-g (colored and prepared for a flannelboard, if desired).

On a flannelboard or table top, place each element of the food chain, one at a time, moving left to right, beginning with Figure 3-13a. As you place each element, discuss how it contributes to the chain. Note: These illustrations may be attached to paper links, which may be joined in order to make a chain, thus reinforcing the idea of a food "chain."

When the chain is complete, discuss what would happen if humans tampered with any part or "link" of the chain. For instance, if humans polluted the air so severely that plants could not grow, and thus must be removed from the chain, what would the butterflies have to eat? If there were no more butterflies (because there is no food for them), what would the dragonflies eat? —and so on until the entire chain has been destroyed.

Food Chain

Figure 3-13a

1. The sun is the source of all food energy. Only plants can use the sun's energy directly.

Adapted from *Ranger Rick's Nature Magazine,* © January 1979, by permission of the publisher, the National Wildlife Federation.

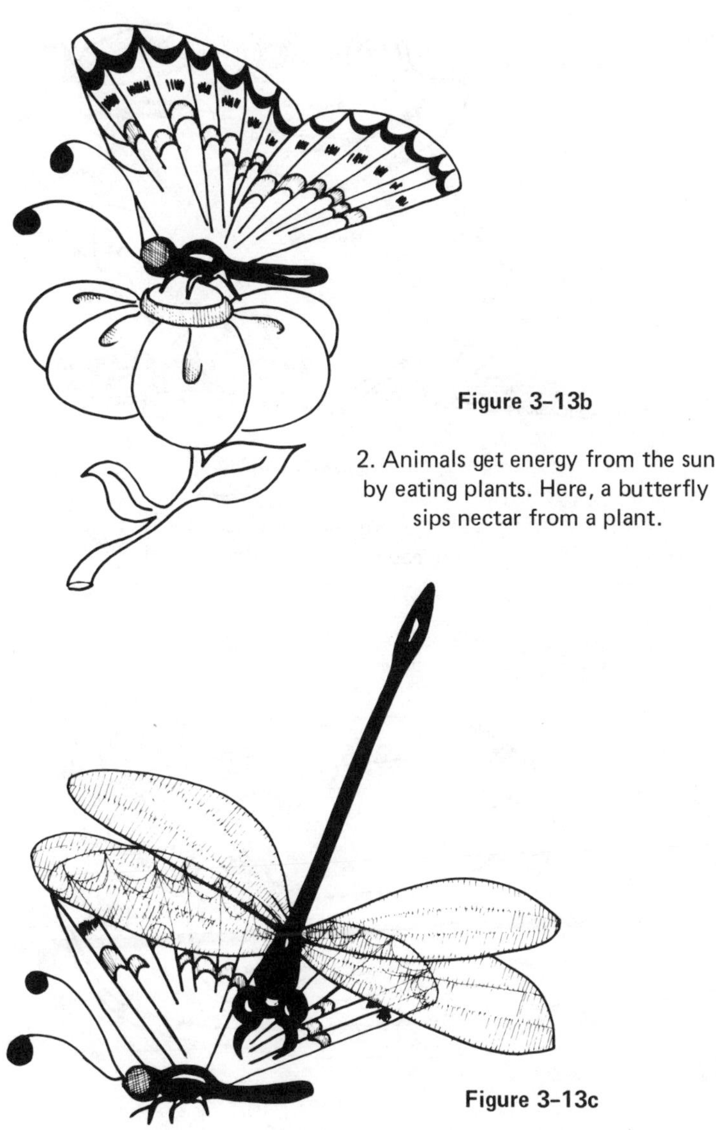

Figure 3–13b

2. Animals get energy from the sun by eating plants. Here, a butterfly sips nectar from a plant.

Figure 3–13c

3. A dragonfly gets food energy by eating the butterfly that got food energy directly from the plant.

Figure 3–13d

4. The energy is then passed on as a frog catches and eats the dragon-fly.

Figure 3–13e

5. Again the energy is passed on as a snake devours the frog.

Figure 3-13f

6. The final link in the chain is the hawk, which gets food energy from eating the snake.

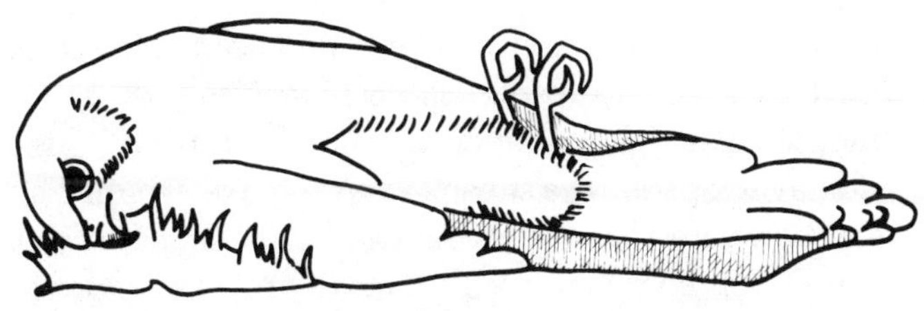

Figure 3-13g

7. The hawk is the final predator that dies. Its body decomposes and helps to enrich the soil, which will grow new plants in order to continue the food chain.

You may wish to explain that if enough plants, and thus animals, in food chains were lost, soon there would be nothing for humans to eat. Humans are the final link in many food chains.

Suggested Questions

1. Why do you think this is called a food chain? Is each plant or animal linked to the other in some way?
2. What happens to the chain if one link is removed?
3. How could the destruction of food chains eventually affect or hurt humans?

Experience: Part III

PURPOSE: To demonstrate one way in which humans interfere with the natural ecological system.

MATERIALS: A shovel; two aluminum pans, one filled with grass sod growing in topsoil and the other filled only with topsoil (potting soil may be used as topsoil); a watering can filled with water.

Go into a natural area that has not been excavated, into an excavated area where topsoil has been replaced, or into a wooded area and remove the top layer of soil (the topsoil). Dig down to the next layer of soil (the subsoil) and remove an equal amount of this soil. Compare the soils, which will differ in color and texture. The topsoil will be richer in color, more brown or more black, and looser in texture. Explain that the topsoil is better for growing plants because it is richer in nutrients that the plants need and because it is looser in texture in order to allow water to flow to the roots and to allow the growth of tiny root hairs. If you wish to grow two identical plants, one in each type of soil, so that the children may observe the superiority of topsoil for growing plants, reserve a small amount of topsoil and a small amount of subsoil for this purpose. Replace the remaining subsoil, topsoil, and grass or leaves (if there were any within the area where you dug).

Outside, have two children hold the aluminum pans filled with topsoil and grass sod at a forty-five degree angle with the ground. These pans represent two hillsides, one that has been stripped of growing plants, grass, and trees and another that is still covered with vegetation. Let one child sprinkle water, representing rain, over the two pans and compare what happens to the topsoil in each pan.

Discuss with the children what is meant by erosion, the wearing away of topsoil. Review the advantages of topsoil for growing plants and, in turn, the disadvantages of erosion. Discuss what happens to the wildlife when humans remove all the vegetation from an area, either on purpose or carelessly, by fire.

Suggested Questions

1. What differences do you see in the topsoil and the subsoil?
2. What happens to topsoil on a hillside that has had all the trees, grass, and plants removed? Would this land be good for growing things?
3. What do you think happens to the animals that live in the woods and grass on the hillside after the woods and grass are removed?

Experience: Part IV

PURPOSE: To learn, in part, what is meant by the term *conservation* and to help the children to understand how they may be conservationists.

MATERIALS: Several earthworms; an earthworm box (a clear plastic storage box with a lid that contains air holes. The box should be filled to two-third's capacity with loamy soil); a cardboard box or cloth large enough to cover the earthworm box;

sand; leaves, grass cuttings, or cornmeal (for feeding the earthworms).

Place the earthworms in the earthworm box and cover the surface of the soil with grass clippings or shredded leaves and a layer of sand. If the box is kept covered (earthworms seek darkness) over a period of several days, the children may observe that the earthworms carry the grass and leaves (enriching the soil) down into the earth—in this case, the box.

Earthworms eat leaves and plant matter on top of the ground. They then burrow into the ground, swallowing soil as they dig. Some of the plant matter and soil that the worm has eaten is digested, and the rest passes through its body. This may be seen at the top of the burrow as "castings." In this process, the texture of the soil is improved as it is ground finer by going through the worm's body, and the soil has been made richer by the material that is excreted from the worm's body. The burrowing process also lets air and water into the ground. This entire process of soil conditioning improves growing conditions for plants.

Explain that conservationists are people who use natural resources wisely or help to improve nature or replace what they use from nature when possible. Therefore, earthworms may be considered natural conservationists, because even though they use the soil for food and a home, they improve it at the same time.

Discuss with the children how they may be conservationists. The following ideas may be helpful in this discussion:

1. Do not waste water. Turn off faucets and use only a small amount of water for showering (showers use less water than baths).
2. Do not waste paper and paper products (writing paper, drawing paper, paper plates, napkins, paper towels). Wasting paper wastes trees.
3. Save paper for recycling.

4. Use reusable products when possible (bottles, dishes, towels, etc.).
5. Collect bottles and cans for recycling.
6. Conserve gasoline by walking or riding a bike instead of riding in a car.
7. Save electricity by turning off lights when not using them and by closing doors when entering or leaving a building.
8. Do not waste food.
9. Plant a garden and use natural ways to control insects (plant marigolds and garlic to ward off insects and place ladybugs or praying mantises in the garden to prey on harmful insects).
10. Plant trees to replenish oxygen and natural beauty.
11. Conserve natural beauty by practicing good outdoor habits: Don't litter, and leave parks, beaches, or other natural and public areas as clean as or cleaner than you found them.
12. Remember that some animals that you may consider undesirable (such as snakes and spiders) may be an important link in a food chain or may be natural predators for rats, mice, or insect pests to keep their populations under control and that they should not necessarily be destroyed.

Suggested Questions

1. Why might the earthworm be considered a conservationist? Does it destroy or harm the soil when using it? Does it improve the soil?
2. How might you be a conservationist? What natural resources could you save or how could you help improve nature?

Follow-Up Activities

Use the food chain pictures from Figures 3-13a-g to create an ecology mobile (see illustration in Figure 3-14). When the

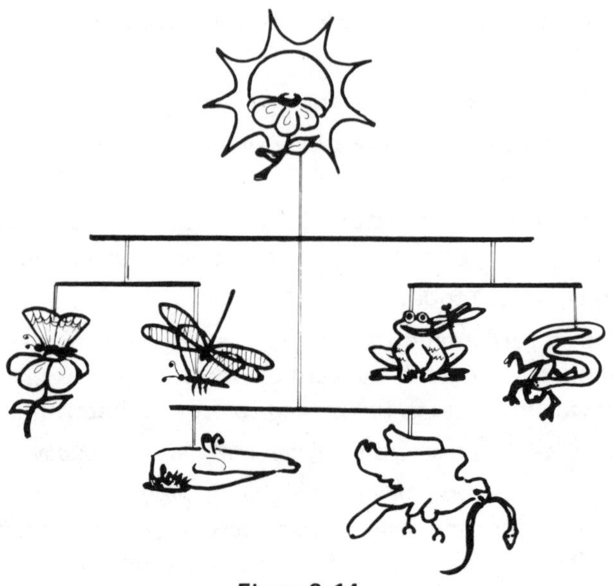

Figure 3–14

mobile is complete, remove one element from the mobile to see what happens to the "balance" of nature.

Play a food sequencing game. Provide pictures of plants and animals that belong to food chains and help the children to place them in the proper group or chain. Then let the children place each element of each chain in the proper order. Some examples of food chains that may be used are the following: grass–field mouse–snake–hawk; grass–deer–wolf; plankton–small fish–larger fish–man; caterpillar–bird–weasel (or cat); ant–beetle–quail–fox; carrot plant–rabbit–fox.

Suggested Readings

Baker, Jeffrey. *Patterns of Nature.* Garden City, N.Y.: Doubleday, 1967.

Collier, Ethel. *Who Goes There in My Garden.* New York: Young Scott Books, 1963.

Conklin, Gladys. *Lucky Ladybugs.* New York: Holiday House, 1968.

Ernst, Kathryn. *Mr. Tamarin's Trees.* New York: Crown Publishing, Inc., 1976.

Gates, Richard. *The True Book of Conservation.* Chicago: Children's Press, 1959.

Hutchins, Ross. *Lives of an Oak Tree.* Chicago: Rand McNally, 1962.

Milgram, Harry. *ABC of Ecology.* New York: Macmillan, 1972.

Moore, Lilian. *Just Right.* New York: Parent's Magazine Press, 1968.

Valens, Evans G., Jr. *Wildfire.* New York: Collins Publishers, 1963.

Wolcott, Patty. *The Forest Fire.* Reading, Mass.: Addison-Wesley, 1974.

Zolotow, Charlotte. *Where the Wind Stops.* New York: Harper & Row, Pub., 1962.

4
SUMMER

INSECTS

Information for Adults

Insects are the most numerous animals in the world, with approximately one million species now identified. Insects may be distinguished from other animals by some of the following characteristics: (1) Insects have hard outer coverings or exoskeletons that are modified to form sense organs and offer protection for the insect. (2) Insects have three main body parts: the head, which supports the eyes, mouth parts, and antennae; the thorax, which supports the legs and wings; and the abdomen, which contains part of the digestive system. (3) Insects have six jointed legs, usually one or two pair of wings, and two antennae. This information is generally true for most insects. Although not all insects will display all of these characteristics, they should display two or more.

Insects probably function more on instinct than by the senses. They do have eyes but probably do not see as well as many other animals. Some insects have simple eyes that probably see little more than light or dark. Some have compound eyes made up of dozens, even thousands of separate lenses. Compound eyes are superior to simple eyes but still provide far from excellent vision.

Only a few insects, such as grasshoppers, crickets, cicadas, and katydids, have real ears. All insects, however, probably rely more on the sense of touch than the sense of sound. Insects use their antennae to smell and feel, and it is these senses that guide them more than the senses of sight or sound.

Insects have mouth parts adapted for the kind of food they eat. These mouth parts are either for biting and chewing or for sucking, usually indicated by a slender tongue or tubelike structure. They feed on plants, living animals, or dead organic matter. Humans generally classify insects as desirable or undesirable by what they eat and whether or not they carry diseases or make pests of themselves. However, insects are generally more helpful than undesirable. They pollinate plants in order to produce food; they produce food, such as honey; and helpful insects feed on those considered undesirable in order to control their population. All insects are important links in food chains, and therefore other life is dependent upon them.

Insects do not have a nose or lungs. Instead, they have a network of breathing tubes that circulate air throughout their bodies. Air enters their bodies through small outside openings.

Most insects reproduce by laying eggs. Some young insects look like mature adults; some go through a change or metamorphosis that includes a wormlike stage. (The Child's World, Inc. P.O. Box 681, Elgin, Illinois, 60120 offers an excellent sequence chart illustrating a complete metamorphosis. The chart is entitled "The Life Cycle of a Monarch.") Some insects go through an incomplete metamorphosis, which means that the young resemble the adult but lack wings. Insects increase in number during the summer, as they reproduce during warm weather and may mate several times during this period. Insects grow by molting or shedding their skins. The inner body becomes too large for the skin, which splits open. The insect then leaves the old covering with a new skin.

Insects are short-lived animals: Generally, they live only a single year or even one season. A few, such as monarch butterflies, ants, termites, cicadas, and some beetles live longer. Most insects still living in the fall are killed by the cold of winter. Some, such as monarch butterflies, migrate; some bees, bugs, and beetles hibernate as adults. Some insects hibernate before reaching an adult form (the cocoon of a moth). Most insects spend the winter as eggs that were laid in the late summer or fall.

Preserved insects offer a good introduction to insect study for those children who may be frightened by live insects. Preserved insects may be obtained from some high school or college biology departments, as students may be required to make insect collections. Insects may often be found lying on walkways, in the grass, in garden areas, on window sills, and near outdoor lights. A painless way to kill and preserve insects is to place them overnight in a freezer. This is not a project to do with children. Preserved insects may be glued in clear plastic boxes or disposable plastic Petri dishes so that delicate body parts will not be easily destroyed.

Insects considered harmful or undesirable because they bite or sting are merely protecting themselves. Insects demonstrate many forms of self-protection. Some escape their enemies merely by flying away; some have extremely hard protective outer coverings. Some insects have a strong odor or a repugnant taste in order to ward off enemies. Many insects use camouflage as protection, matching or blending into their surroundings so that they may not easily be seen. Some harmless insects mimic the appearance of stinging insects in order to fool their enemies. Certain insects, such as beetles, "play possum" if caught, so that they appear to be dead.

In presenting this unit, remind the children that although insects show a very low form of intelligence, they can feel pain, should be given food and water, and should not be tortured. For ecological reasons, return live insects to their natural habitats within a few days after their capture. The one exception to this rule is ants, which may stay for weeks in an ant farm but should eventually be released. Information on specific insects, their characteristics, and habits may be found in Appendix III. Many issues of *Ranger Rick's Nature Magazine*, published by the National Wildlife Federation, offer excellent sources of pictures and information on how insects protect themselves.

Experience: Part I

PURPOSE: To learn how insects are distinguished from other animals.

MATERIALS: Preserved insects; a magnifying glass; pictures of insects and other animals.

Discuss with the children the characteristics that help to distinguish an insect from other animals. Use the preserved insects to point out these features. As the children examine the preserved insects, suggest that they find the legs, wings, three body parts, antennae, mouth parts, eyes, ears (if there are any), and so on.

Display the pictures of the insects and the other animals. Let the children categorize the pictures into two groups: insects and non-insects. Use the preserved insects and the pictures to discuss with the children how the mouth parts of insects are adapted for what they eat and the methods of protection insects demonstrate.

Suggested Questions

1. Can you find six legs, two antennae, and three body parts on this animal? Is this an insect or another type of animal?
2. This insect has large jaws. How does it eat? This insect has a long slender tongue. How does it eat?
3. This animal has eight legs. Is it an insect?
4. This animal has two wings and two legs. Is it an insect?
5. This animal has six legs and two wings. Is it an insect?

Experience: Part II

PURPOSE: To observe live insects—their characteristics and habits.

MATERIALS: A "critter cage" (a temporary home made of a stiff small gauge wire, twelve to eighteen inches in width, from the hardware store, and two metal cake pans); food to attract insects (raw meat, dead insects, sugar, honey, small pieces of fruit); a small jar with a lid or a net.

The children may participate in making a "critter cage." Cut the wire so that, when rolled into a tube, its circumference will match that of the inside of the cake pans. Wire the tube together and place a cake pan on each end. When set upright, the top cake pan becomes a lid. A small limb may be placed inside the cage so that insects will have a place to rest and hide (see Figure 4-1).

Figure 4-1
Critter Cage

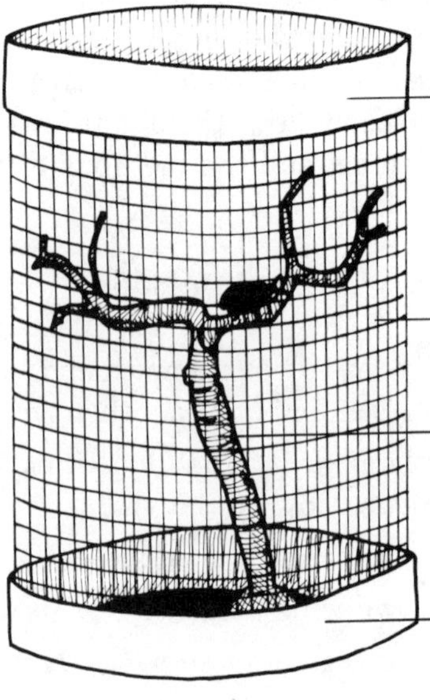

Set out a variety of food to attract live insects and observe what sort of insect is attracted to each type of food. Capture insects with the jar or net and place one or more insects in the cage for observation. Specific information for the discussion of certain insects is given in Appendix III, and the suggested questions given at the end of this experience will help to direct the children's observations.

The children might enjoy performing some of the following activities with their captured live insects.

Test in order to see how well an insect can see or sense the children's presence by letting the children, one at a time, move slowly and quietly near the cage. How close are they able to get before the insect becomes frightened or moves away?

Let the children test the insect's hearing ability by standing about two feet away from the cage and clapping loudly. Does the insect respond in any way to the sound?

Measure the distance a hopping insect is able to travel in one jump. How many hops does it take for the insect to travel one yard or one meter? Place the insect in the refrigerator for five minutes. Does this affect its activity level?

Suggested Questions

1. How do you know this is an insect?
2. What kind of food does this insect eat? What sort of mouth does it have?
3. Does this insect ever clean itself? How?
4. Does the insect move its antennae much? Why do you think it does this?
5. Can this insect walk up the side of the cage or limb? Can it walk upside down?
6. How might this insect protect itself?
7. Does this insect make any noise? What sort of noise? When does it make the noise?

Follow-Up Activities

Let the children make "bug jugs" for catching and temporarily keeping their own insects. You'll need gutter wire (from a hardware store or ordered from a Sears catalog) and two hair spray can tops. Cut the gutter wire long enough so that when the wire is rolled into a tube and wired together, the ends of the tube fit very tightly into the hair spray can tops. One hair spray can top becomes the lid, the other the bottom. A wire or pipe cleaner handle may be attached for carrying, if desired (see Figure 4-2).

As an art project, let the children make insect models with vegetables and pipe cleaners. A stalk of celery becomes a praying mantis with pipe cleaners for legs; potatoes become beetles; and carrots sliced crosswise into sections and joined by toothpicks become caterpillars.

Set up a commercial ant farm or make your own with a commercial-sized clear glass jar. Place a quart size jar upside down in the large jar, so that ants may not dig tunnels in the center where they would not be observable. Carefully dig an ant bed, trying to reach the eggs, which look like grains of uncooked white rice. Place thirty to fifty ants and some soil into a plastic bag; tie the bag and place it in a refrigerator for about two hours

Figure 4-2
Bug Jug

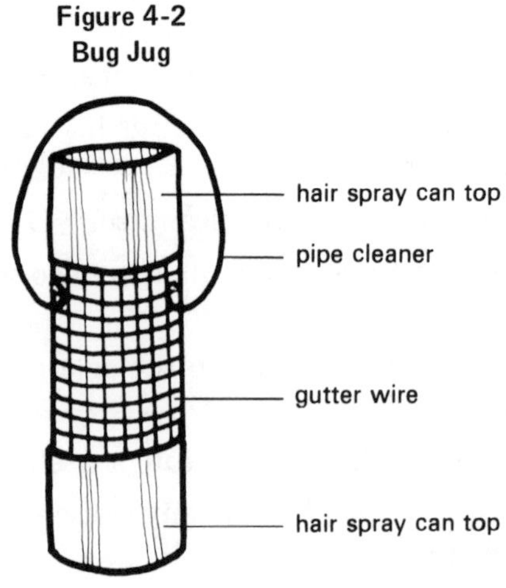

in order to make the ants less active and easier to handle. Put the ants and soil into the large jar. Drill air holes in the lid of the jar and fill the holes with cotton so the ants may not escape. Keep the ant farm moist, not wet, and covered, except when observing it. Place food, such as fruit, sugar, honey, or crumbs, on top of the soil in the jar about once a week.

Kits are available that include everything needed to raise butterflies. For a catalogue, write to Insect Lore Products, P.O. Box 1535, Shafter, California, 93263.

Let the children use their imaginations to design a new insect or to combine parts of insects in order to create a new breed. Remind the children that, in order to qualify as an insect, this animal must have three body parts and six legs.

As the children bring in insects they have caught, set up an "insect zoo." Provide the insects with clear glass jar cages provided with air holes. The insects in the zoo must be temporary residents, as they should be released within two days.

Suggested Readings

Carle, Eric. *The Very Hungry Caterpillar.* New York: Scholastic Book Service, 1974.

Caudill, Rebecca. *A Pocketful of Cricket.* New York: Holt, Rinehart & Winston, 1964.

Conklin, Gladys. *Lucky Ladybugs.* New York: Holiday House, 1968.

——. *We Like Bugs.* New York: Holiday House, 1962.

Goudey, Alice E. *Butterfly Time.* New York: Scribner's, 1964.

Hitte, Kathryn. *Boy, Was I Mad.* New York: Parent's Magazine Press, 1969.

Huntington, Harriet E. *Let's Look at Insects.* Garden City, N.Y.: Doubleday, 1969. (Excellent pictures.)

Hutchins, Ross. *Hop, Skim, and Fly.* New York: Parent's Magazine Press, 1970.

Martin, Bill, Jr. *Ten Little Caterpillars.* New York: Holt, Rinehart & Winston, 1967.

Mizmura, Kazue. *The Way of an Ant.* New York: Thomas Y. Crowell, 1970.

Mousdale, John. *Butterflies.* New York: Wonder Books, 1973.

Kepes, Juliet. *Lady Bird, Quickly.* Boston: Little, Brown, 1964.

Podendorf, Illa. *The True Book of Insects.* Chicago: Children's Press, 1954.

Selsam, Millicent E. *Questions and Answers About Ants.* New York: Scholastic Book Services, 1967.

Selsam, Millicent, and Hunt, Joyce. *A First Look at Insects.* New York: Walker and Co., 1974.

SPIDERS

Information for Adults

There are thousands of different kinds of spiders that have been identified. Although many people are frightened by spiders and think of them as pests, they are extremely helpful in that they consume large amounts of plant-eating insects. Seldom do spiders harm humans, although they are able to bite as a means of protection. The only spiders in the United States able to inflict harmful poisoning are the black widow and the brown recluse. Their bites can be fatal but seldom are.

Spiders are not insects: They are members of the order Araneae: and some of their relatives include crabs, lobsters, shrimp, crayfish, scorpions, and daddy longlegs. Spiders may be distinguished from other animals in the following ways: (1) Spiders have two main body parts, the thorax and the abdomen. (2) The body is encased in a hard exoskeleton. (3) Spiders have eight legs and no wings or antennae. (4) Spiders generally have more than eight eyes (usually eight simple eyes). (5) They have no chewing jaws but are equipped with a pair of fangs for paralyzing prey. Their fangs are extensions of the chelicerae, or nippers, which are used for crushing prey.

Not all spiders spin webs. Some, such as crab spiders, wolf spiders, or jumping spiders, wander around in search of prey or sit in wait for prey to come to them. Spiders that spin webs do so in order to capture food. Webs are spun from an extremely strong silk that comes from spinnerettes (usually six) on the spider's abdomen. The silk comes from glands within the abdomen. Webs come in many shapes and patterns, including orb webs, funnel webs, sheet webs, and mesh webs, which are masses of silk. Webs begin with foundation lines or the framework. Then sticky lines are added for the purpose of snaring prey. The spider knows which part of the web is made of dry silk and which part is sticky, so it is able to walk on the web without getting stuck.

When a web has been completed, the spider may attach a final line that acts as a bridge leading from the web. The spider then moves to a nearby hiding place (usually under a leaf). Movement of the line leading to the web alerts the spider that prey has been captured. Most spiders remain in the web while waiting for captured prey. Once the prey is captured, the spider wraps it in a silk covering and paralyzes it with its fangs. The spider then sucks the liquids from the victim's body or saves it for a later meal. Each time prey is caught, the web is damaged and must be repaired. A spider spins a new web almost every day.

A spider reproduces by laying eggs in a silken sac that is often about the size of the adult's body. The silk covering protects the eggs or the small spiderlings, which sometimes hatch and stay inside the sac for the entire winter. Some spiders lay eggs in the fall, and the spiderlings hatch in the spring. Some spiders lay eggs throughout the warmer months of the year. The mother spider may die shortly after the eggs are laid, forcing the spiderlings to fend for themselves. The young spiders look like miniature adults and grow by molting or shedding their skins.

Experience: Part I

PURPOSE: To learn how a spider may be distinguished from other animals.

MATERIALS: A large picture of a spider (the books suggested at the end of this unit offer excellent pictures); small individual pictures of spiders, insects, and other animals.

Discuss with the children the distinguishing characteristics of spiders (see "Information for Adults") and point out these characteristics in the pictures of the spiders.

Place the individual pictures of spiders, insects, and other animals on a table and let the children classify the pictures into two groups: spiders and non-spiders.

Suggested Questions

1. How many legs does a spider have? How many body parts? Does a spider have wings or antennae?
2. How many legs does this animal have? How many body parts? Is this a spider?
3. This animal has eight legs and one body part. Is it a spider?

Experience: Part II

PURPOSE: To observe a spider spinning a web and capturing prey.

MATERIALS: A live web-spinning spider, such as a garden spider or other orb weaver (spiders may be easily caught and trapped with a jar and a lid); a "spider box" (this consists of a cardboard box approximately 12″ × 15″ × 8″ or larger painted black with a door cut in the back, a small limb to fit inside the box, clear plastic wrap, and masking tape).

Place the small limb inside the black box: This will give the spider something on which to attach the web. Cover the open side of the box with clear plastic wrap and seal it on all four sides with masking tape. Place a small container of water in the box. (In nature, spiders obtain their water from dew on plants.) Place the live spider in the box and seal the back opening with masking tape.

Spiders generally spin webs in the early morning and at twilight. When a web has been completed, release a live insect through the opening at the back of the box (houseflies, beetles, and moths are easily obtained). This will allow the children to observe the spider capture its prey. The spider will need to be fed only every few days. You may wish to let the children experiment with placing other types of food in the spider box so that they may discover what a spider will and will not eat.

While observing the spider, discuss such details as where the spider gets the silk to spin the web, why the spider does not stick to its own web, how the spider eats, the purpose for which the spider uses its web, and so on.

Suggested Questions

1. Why does a spider spin a web?
2. From where does the silk seem to be coming?
3. How does the spider eat? What does it eat?
4. Why does the spider not stick to its own web?

Experience: Part III

PURPOSE: To closely examine webs by "catching" them.

MATERIALS: Black construction paper (one piece for each web to be caught); white spray paint; hair spray; one or more spider webs.

Locate a spider web and carefully remove the spider. Let the children spray the dry web with white paint. Allow the paint to dry, and then spray the web lightly with hair spray. While the web is sticky from the hair spray, place the black construction paper against the web and gently release the main strands of the web. The web should remain intact and stick to the black paper so that the children may examine its construction. Various types of webs may be caught so their designs may be compared. Note: Although spiders build new webs almost every day, this activity should not be "overdone." Collect only enough webs to compare the construction of different types of webs.

Suggested Questions

1. In what shape is this web?
2. What does the web seem to be made of?
3. Do all the webs look alike?
4. How would you go about making a web? What would you do first?

Follow-Up Activities

Let the children design a new twentieth-century spider. What does it look like? How does it capture food? What must it have to qualify as a spider (how many legs and body parts)?

Let the children attempt to make a spider web. Provide them with string or thread to use as silk. They will have more respect for the spider's ability to make a web after they have tried to make one.

Suggested Readings

Dupre, Ramond S. *Spiders.* Chicago: Follett Publishing Co., 1967.

Graham, Margaret B. *Be Nice to Spiders.* New York: Scholastic Book Services, 1967.

Lexau, Joan M. *The Spider Makes a Web.* New York: Scholastic Book Services, 1979.

Morris, Dean. *Spiders.* Milwaukee, Wis.: Raintree Children's Books, 1977.

Shuttlesworth, Dorothy. *The Story of Spiders.* Garden City, N.Y.: Doubleday, 1959. (Excellent pictures.)

White, E. B. *Charlotte's Web.* New York: Dell Pub. Co., Inc., 1952.

THE OCEAN

Information for Adults

There is more water than land covering the face of the earth. The larger bodies of salt water are called oceans, and the smaller bodies of salt water are called seas. Both are salty because, over the years, salts contained in the soil have washed from the land and accumulated in the oceans and seas. The older a body of water is, the saltier it is, because more salt has had time to accumulate. Salt is also washed from the land into rivers and lakes, but since most of these are not as old as oceans and seas, there has not been a significant accumulation of salt.

The ocean appears to be different colors in different places and at different times, a fact attributed to several elements. Color is, in part, created by microscopic plants, called algae, that are in the water. These are usually blue in color, making the water blue in appearance. Some algae are green, creating a green appearance, or even red, giving a red color to the water.

The amount of sunlight shining on a body of water can also change the appearance of the color. On a sunny day, the water appears lighter and brighter because it reflects the bright blue of the sky, and on an overcast day, the water reflects the dull grey color of the sky. The color of the sand underneath the water can also create color differences. Sand can be any variety of colors, ranging from almost pure white to black or slightly yellow to orange.

Sand is found under and near the ocean because the constant motion of the water grinds pieces of rock together to create grains of sand. The sandy bottom of the ocean is called the floor, and the sand at the water's edge is called the shore or beach. Without its water covering, the ocean floor would look much like the land, with mountains, valleys, rocks, and plants. Sand at the ocean's edge is constantly being shifted about by the movement of the water, so the shoreline is in a constant state of change.

Waves in the ocean are created by wind: Stronger winds create larger waves, and milder winds create smaller waves, sometimes ripples. A wave can rise as high in feet as one-half the speed of the wind in miles per hour; therefore, in a 40 mph wind, a wave can rise as high as twenty feet. The height of a wave is measured from crest (the highest point) to trough (the lowest point).

A wave needs time and space to build up; thus, the larger the body of water and the deeper the water, the higher a wave can become. The farther from land a wave develops, the larger it can become, since it has more time to build before meeting the shore. Waves along the shoreline are larger when the wind blows toward the shore.

Waves move at different speeds, according to conditions, but always in regular patterns of alternating large and small waves. Wave movement under the surface of the water is always in a circular motion, so that waves, in a sense, do "roll" onto the beach.

Tides are quite different from waves and are caused by the magnetic attraction of the moon and sun, both of which are able to pull water away from the earth and towards themselves. The larger the body of water, the greater the effect the sun and moon can have on it. Even though the moon is smaller than the sun, it is nearer the earth and thus has more gravitational pull and more influence on the tides.

Whenever the earth, moon, and sun are in a direct line

with each other, the moon and sun together exert a great gravitational pull, and the high tides (called spring tides at this time) will be very high. When the sun and moon do not pull together, smaller tides (neap tides) are created. High tide occurs in a particular area when the moon's gravitational pull parallels an imaginary line running from the moon and directly through the center of the earth. The area on the side of the earth nearest the moon will have a higher high tide than the area on the exact opposite side of the earth because the nearer the moon is to a particular spot, the greater the pull it exerts on that spot. The parts of the earth one-quarter turn away from the moon are not in the gravitational pull of the imaginary line from the moon through the center of the earth and thus will have low tide.

Because of the rotation of the earth, there are two high tides and two low tides each day, although in some places, only one high tide and one low tide occur daily. The moon's movement creates a change in its position each day. This causes the two high tides and the two low tides to be not exactly twelve hours apart but about twelve and one-half hours apart.

Experience: Part I

PURPOSE: To observe how salt can wash from the land into the ocean, causing the ocean to be salty.

MATERIALS: Table salt mixed with clean sand in about equal parts; a cookie sheet or an aluminum pan; a pan or a bowl of water; a sprinkling can filled with water.

The adult should begin this experience by explaining to the children that the soil contains salts. Have the children taste the water in the pan or bowl in order to see whether it is salty. Now pour about one cup of the salt and sand mixture onto the cookie sheet or aluminum pan to represent the land and the salts in the soil, and then hold the cookie sheet against the pan of water at an angle (the way land slopes toward the ocean). Ask a child to sprinkle water to represent rain over the cookie sheet and observe what happens. After some of the salt and sand have washed into the water, ask the children to taste the water again to see if it is salty. Let a child sprinkle more water over the cookie sheet and suggest that this water represents rainfall for another day or year. This activity may be repeated several times until all the salt and sand have washed into the pan of water. Have the children taste the water after each sprinkling in order to determine whether the water gets saltier as each day or year of rainfall is represented.

Suggested Questions

1. What happens to the salt on the "land" or cookie sheet as the "rain" washes over it?
2. How does the water change when the salt is washed into it?
3. What happens to the water as more salt is washed into it over many years?
4. How does the ocean taste?
5. Do you think that salty oceans have had a lot of salt wash into them over many years?
6. Why are rivers and lakes not salty?

Experience: Part II

PURPOSE: To add coloring to water to observe that water changes appearance according to what it contains.

To learn that there are plants in the oceans and seas that change the color of these bodies of water.

MATERIALS: Two glasses of water; one bottle of green food coloring; one bottle of blue food coloring.

Let a child drop four or five drops of green food coloring into one glass of water and four or five drops of blue food coloring into the other glass of water. Discuss whether the glasses of water now look alike or if they are different in color and what the reason is for the difference. Tell the children that plants so small that they cannot be seen without a microscope are in the water and see if the children are able to explain how these plants could create color differences.

Suggested Questions

1. Why are the glasses of water different colors after food coloring is added?
2. If tiny plants in the ocean water have different colors, how would they change the color of the water?

Experience: Part III

PURPOSE: To look through different depths of water in order to determine whether depth affects color.

MATERIALS: The glass of blue water from the previous experience; one sheet of white paper; one empty glass; a ruler (optional).

Pour about one and one-half inches of blue water from the full glass into the empty glass while the children observe in order to establish that both glasses now contain blue water. Try to get the children to think of one glass of water as deep and the other as shallow. They may use the ruler to measure the depth of the water in the glasses in order to determine that one glass contains more water than the other. Place the glasses of water on a sheet of white paper in order to make color contrasts more visible. Have the children stand directly over the glasses and look down into the water in order to determine whether the color of the water in the two glasses appears different or the same. Ask which glass of water appears darker in color and which appears lighter.

Suggested Questions

1. What color is the water in both glasses?
2. Do the glasses contain the same amount or different amounts of water?
3. Which glass of water is deep and which is shallow?
4. When you look directly down into the two glasses of water, how is the color different? Which glass is dark blue and which is light blue?
5. How do deep water and shallow water in the ocean differ in appearance?

Experience: Part IV

PURPOSE: To observe that the color of water changes in appearance according to how much light is shining on it.

MATERIALS: Two glasses of blue water from the previous experience; a ruler (optional); a source of bright light (a lamp or a sunny window).

Pour blue water from one glass into the other glass until the water level in both glasses is the same. Measure the levels with the ruler or establish visually that both glasses contain the same amount of blue water. Have the children put one glass in a rather dark or shaded place in the room and the other glass in the source of bright light. Discuss whether the colors now appear alike or different, and then ask what causes the color difference.

Suggested Questions

1. Do both glasses have the same amount of water?
2. What color is the water in both glasses?
3. After separating the glasses, ask, "Which one is light blue in color and which one is dark blue in color?"

4. Where is the glass of light blue water sitting? Where is the glass of dark blue water sitting?
5. What, in nature, causes light to shine on water?
6. Why does the ocean look dark on a cloudy day?

Experience: Part V

PURPOSE: To observe that the color of water changes in appearance according to the color of the surface under the water.

MATERIALS: Two glasses of blue water from the previous experience; one piece each of white and black construction paper; a ruler (optional).

Again establish that both glasses contain equal amounts of blue water. Discuss with the children that sand may be a variety of colors. Then ask them to think of the pieces of paper as sand under the water and place one glass of water on each piece of paper. Discuss whether the color of the water appears to change in any way and how it differs. Let the children give their ideas of why the color of the water changed. They may like to experiment by swapping the two glasses or by putting different colors of paper under the water.

Suggested Questions

1. Do both glasses have the same amount of water? Is the water in both glasses the same color?
2. How did the color of the water change when the glasses were placed on paper of different colors?
3. Which one is light in color (on what color paper is it sitting)? Which one is dark in color (on what color paper is it sitting)?
4. How can the sand under the ocean change or affect the color of the water?

Experience: Part VI

PURPOSE: To make waves by blowing air (which represents wind) over a pan of water (which represents the ocean).

To observe that larger waves form if there is a great deal of wind and that smaller waves form if there is only a little wind.

To observe that larger waves can form if water is deep.

MATERIALS: A pan (9″ × 12″ or larger) containing about one-half inch of water; a pitcher or container of water; a ruler.

Ask the children to look at the pan of water in order to determine whether there are any "waves." Ask them to make some wind over the surface of the water by blowing gently, and then see if there are any waves. Now let them make a strong wind by blowing very hard over the surface of the water and observe what happens to the waves this time.

Let the children watch as you add more water to the pan. Ask them to measure before and after adding more water in order to establish that the water in the pan is now deeper. Again have the children blow over the pan of water and discuss whether the waves became larger or higher in the deeper water.

Suggested Questions

1. What caused waves to be in the water?
2. What happened to the waves when you blew harder to make more wind?
3. What happened to the waves when we made the water in the pan deeper?
4. On what kind of day would we see big waves in the ocean?
5. In the ocean, would we see very big waves in deep or shallow water?

Experience: Part VII

PURPOSE: To learn that the moon can pull water upward much like a magnet can pull metal.

To learn that high tide occurs when the moon is pulling the ocean and that low tide occurs when the moon is not pulling the ocean.

MATERIALS: A strong magnet; a light metal object that the magnet will attract, such as a paper clip; a lightweight scarf.

Tides are difficult to explain to children, and younger children may not even be aware that they exist; therefore, the adult may wish to reserve this activity for older children or for children with a special interest in the subject. If the children have not had previous experiences with magnets, let them experiment in order to see that the magnet pulls or picks up certain objects.

Explain that the moon can pull water much like a giant magnet. Place the paper clip on the table top with the scarf covering it and ask the children to think of the scarf as the ocean and the magnet as the moon. Let them test to see what happens as the magnet (the moon) comes closer to the metal object under the scarf (the ocean). Explain that high tide occurs when the water is pulled upward by the moon and that low tide occurs when the moon releases the water. Be sure that the children understand that there is no metal object in the ocean by which the moon pulls the water.

Suggested Questions

1. What happens to the water (the scarf) as the moon (the magnet) comes close to it? What happens to the water as the moon moves away from it?
2. What happens to the water level at the shoreline when it is high tide? when it is low tide?

Follow-Up Activities

While displaying pictures of the ocean, discuss some questions, such as the following: Is it a windy day or a calm day? How can you tell by looking at the water? Why is the water blue? Green? Can you find some deep spots in the water? How do you know these are deep spots? Where is the water shallow? How do you know it is shallow there? Is the sun shining brightly on the water? How does this affect the color of the water?

If it is not possible to take a field trip to the ocean, let the children make an "ocean" and a "beach" by adding water to the sandbox. Suggest that they build a sand castle. See if they are able to think of some ways to use their various senses in order to explore the ocean.

Let the children design an ocean city. How would the people live under the water? How would they dress? How would they get food? What would their homes look like? What sort of transportation would they use? What kind of pets would they have?

Suggested Readings

Carrick, Carol, and Carrick, Donald. *Lost in the Storm.* New York: Seabury Press, 1974.

Carter, Catherine. *The True Book of Oceans.* Chicago: Children's Press, 1958.

Cartwright, Sally. *The Tide.* New York: Coward, McCann & Geoghegan, 1970.

Goldin, Augusta. *The Sunlit Sea.* New York: Thomas Y. Crowell, 1968.

Greenhood, David. *Watch the Tides.* New York: Holiday House, 1961.

Kinney, Jean, and Kinney, Cle. *What Does the Tide Do?* New York: Young Scott Books, 1966.

Lund, Herold. *The Paint-Box Sea.* New York: McGraw-Hill, 1973.

McCloskey, Robert. *A Time of Wonder.* New York: Viking, 1957.

FISH

Information for Adults

Fish are animals with backbones and are able to live only in the water. A fish has gills instead of lungs and a tail and fins instead of legs and feet. Gills are found on the head or neck just behind the eyes, and most fish (excepting sharks and rays) have gill covers. A fish opens its mouth and takes in water, which flows through the gills. Blood vessels in the gills remove oxygen from the water, and the remaining water flows out as the fish opens its gill covers.

A fish wriggles its body and tail in order to propel itself through the water. Fins are used to guide the fish, to hold it upright in the water, to stop, and to change directions. A fish has many fins on its body, each designed for its type of work. The tail of a fish may be compared to the oars of a boat, and the fins may be compared to the rudder.

Some fish are able to adjust their position in the water from surface to bottom through the use of an air bladder. Much like a balloon inside the fish's body, the bladder can fill with air, causing the fish to rise toward the surface of the water, or empty of air, causing the fish to sink toward the bottom. A fish without an air bladder will sink unless it is swimming.

The body of most fish is covered with scales, which grow from the skin outward. The scales on most fish overlap, much like shingles on the roof of a house; some scales fit side by side like the tiles of a floor; and some scales (like those on rays and sharks) are like small teeth outside the body. Scales grow larger as the fish grows; and on most scales, growth rings, much like those inside a tree trunk, may be found. These rings may be counted to determine the age of the fish. Scales offer some protection for a fish, but they are not an outside skeleton like the hard outer covering of a mollusk. Most fish, called bony fish, have a hard inner skeleton, but sharks and rays have inner skeletons made of cartilage much like that found in the human nose.

A fish eats in much the same way it breathes, by taking

water and food inside its mouth and retaining the food while releasing the water through the gills. Most fish have eyes but no eyelids. There are no visible ear openings on the outside of a fish's body. Fish do have earlike structures under the skin, but they probably do not hear very well. It is believed that perhaps fish use a lateral line (a pressure-sensitive structure within their bodies) more than a sense of hearing and that they are able to detect changes in pressure rather than specific sounds. Fish have a keen sense of smell, and their nostrils, located on the face rather than on a noselike structure, are used solely for smelling and not for breathing.

Fish are cold-blooded animals—that is, the temperature of their bodies is the same as that of the water surrounding them. Fish do feel changes in water temperature, but they prefer water of a particular temperature. Therefore, some fish may seek warmer water in the winter; some merely move closer to the bottom where the water stays warmer; and some migrate great distances.

All fish reproduce from eggs. Some lay the eggs; others, such as sharks and guppies, carry the eggs inside them and bear their young alive. Some fish lay eggs encased in a sticky substance that attaches to water plants or falls to the bottom; some build nests for their eggs; and some carry their eggs in pouches or even in their mouths.

Fish have many ways of protecting themselves. Some fish are very fast swimmers and can outrun their predators. These fast swimmers will have streamlined bodies, and generally, the more streamlined the body, the faster the fish is able to swim. Some fish hide in sea plants or in piles of rocks. Fish that hide may also be equipped with some sort of camouflage, matching their surroundings either in color, pattern, or texture. Many fish have weapons for protection; sharp teeth, pointed snouts, spines, stingers, even poison. Fish may use tricks as a means of protection: The puffer fish or the blow fish can fill itself with air so that it appears to be many times its normal size; some harmless fish resemble dangerous fish in order to confuse their predators. Many types of fish live in groups called schools; the large numbers offer protection for the individual.

When discussing fish, the adult may wish to clarify which animals living in the water are fish and which are not fish. Some other water animals that the children will be familiar with that are not fish are shrimp, clams, octopuses, squids, sand dollars, star fish, sea urchins, jelly fish, sea snakes, tadpoles, dolphins, porpoises, and whales. Sharks, rays, eels, and sea horses are fish.

The National Wildlife Federation, 1412 16th Street, N.W., Washington, D.C., 20036 offers a variation of the card game "Old Maid" that has pictures of many fish of North America. MAB Graphic Productions Inc., 5 Kern Street, West Babylon, New York, 11704 offers flash cards with excellent pictures of fish.

Experience: Part I

PURPOSE: To locate and name some familiar parts of a fish.

To locate and learn the name for some unfamiliar parts of a fish.

To examine a scale from a fish closely.

MATERIALS: A whole fish (complete with head); a strong magnifying glass or a microscope.

Let the children examine the fish, locating and naming some of the parts that may be familiar to them. Suggest that they try to find eyes, eyelids, mouth, teeth, nose, nostrils (external nasal openings), ears, and tail. Now give them the names for and let the children examine some of the more unfamiliar parts of a fish, such as fins, gills, and scales.

Remove a scale from the fish and examine it closely under the strong magnifying glass or the microscope. If possible, count the growth rings in order to determine the age of the fish. Notice how all of the scales of the fish are arranged on its body.

Suggested Questions

1. Do the gill covers open and close?
2. How do the fins move?
3. How many fins does the fish have?
4. Does this fish have many scales? Do the scales overlap on the fish's body?
5. Why do you think fish have scales?

Experience: Part II

PURPOSE: To observe some habits and characteristics of a live fish.

MATERIALS: One or more live fish in a bowl or aquarium.

Let the children observe the live fish swimming, breathing, eating, and resting. Then offer some of the following questions for thought.

Suggested Questions

1. How does this fish move itself through the water?
2. How does the fish use its tail and fins?
3. Why do you think the gill covers are opening and closing?
4. Where does the fish go to rest? Does the fish lie down to rest?
5. What shape is the fish's body—streamlined, balloon-shaped, a question mark shape?
6. Does this fish eat its food near the surface or near the bottom of the water?

Experience: Part III

PURPOSE: To learn which animals living in the water are fish and which are not fish.

To learn some ways fish protect themselves.

MATERIALS: Pictures of various types of fish, of sea animals that are not fish, and of other animals; two small boxes.

Compare the pictures of the fish and those of other animals and discuss what makes an animal a fish or not a fish. The pictures may be used to discuss some reasons for differences in appearance among fish—color, adaptations, shape of body. Let the children use the two boxes to categorize the pictures into fish and other animals.

Using the pictures of the fish, discuss some ways that fish have of protecting themselves. Let the children categorize the pictures into groups according to the way the fish protects itself. Some of the groups may be protected through fast swimming, through tricks, through living in schools, or through hiding (including camouflage).

Suggested Questions

1. What makes an animal a fish?
2. With individual pictures, ask, "Is this a fish? Does it have gills, fins, and scales? Does it live in the water? Can it swim?"
3. How can you tell if a fish is a fast swimmer? Can you find a picture of a fish that is a fast swimmer?
4. What might this fish do to protect itself from an enemy?

Experience: Part IV

PURPOSE: To understand, in part, how a fish uses its bladder in order to change positions or depth in the water.

MATERIALS: Two balloons, one containing water and partially inflated, the other filled with water but not inflated; an aquarium or a large bowl of water.

Let the children watch the preparation of the balloons so that they will be aware that although both are filled with water, one contains air and one does not. Place both balloons in the tank of water and observe that one sinks and one (partially inflated) floats or stays near the surface. Tell the children that a fish has something called an air bladder that is like a balloon inside its body. See if they are able to suggest how a fish may use its air bladder to change its position in the water.

Suggested Questions

1. Which balloon floated and which one sank?
2. How could a fish use its air bladder, which is sort of like a balloon, to help it sink to the bottom of the water or rise to the top?

Experience: Part V

PURPOSE: To understand, in part, how a fish is able to breathe under water.

MATERIALS: A drinking straw; a fish tank with a bubbler; a fish bowl containing fish and water that has not been recently changed (optional).

Begin by establishing that air, which contains oxygen, comes out of one end of the drinking straw when someone blows through the other end. Let a child use the straw to blow air bubbles into the tank. Establish that other air bubbles may be seen in the tank and that there must be air, and thus oxygen, in the water. Explain how the fish might take water and air into its body and use the oxygen but release the water.

If a bowl of fish (without a bubbler) containing water that has not been changed recently is available, this may be used to demonstrate what fish can do when there is not enough oxygen in the water. Let the children watch the fish come to the surface for air. Explain that the oxygen has been used up in the old water, and then see if the children are able to suggest why the fish are coming to the surface.

Suggested Questions

1. What comes out of the straw when we blow air through it?
2. What did the air from the straw make in the water?
3. Do you see other bubbles in the water?
4. How is the fish able to get the oxygen from the water?
5. Why does the fish go to the surface of the water when the water in the bowl is not fresh?

Experience: Part VI

PURPOSE: To see how fish are able to live in water of various temperatures.

MATERIALS: A fish tank containing nontropical fish, such as goldfish.

Discuss what is meant by "cold-blooded." Turn off the heater in the fish tank and float a container filled with ice cubes on the surface of the water. Observe what happens to the fish as the water gets colder. It may be necessary for the children to feel the water in the tank in order for them to see that the water at the surface, near the ice cubes, is much colder than the water at the bottom of the tank. Suggest that fish living in lakes, seas, oceans, and rivers might move to the bottom of the water away from the cold air on the surface. Discuss migration in terms of another way a fish could find warmer water in the winter.

Note: Do not allow the ice to remain in the tank for an extended period of time, as extreme cold could harm the fish.

Suggested Questions

1. What happens to the water around the ice cubes?
2. Where do the fish go when the water near the top of the tank gets cold?
3. Do the fish move around as much in the cold water?
4. What else could fish do to find warmer water in the winter?

Follow-Up Activities

Discuss ways in which fish are helpful or may be used by humans. Discuss ways that humans can be helpful to fish.

Fish may be trained by feeding them from the same corner of the tank every day and blowing a whistle just before feeding time. In a few days, they will come to their feeding corner whenever the whistle is blown.

For a creative movement activity, let the children mimic the movements of a fish swimming, eating, or escaping from an enemy. Let them tell how they think their lives would be different if they had a tail instead of feet or gills instead of lungs.

Let the children draw an underwater scene, complete with fish. Use watered tempera to paint the background and strips of tissue paper to represent underwater plants. Let the children dip the tips of their fingers in a dark colored tempera paint and make prints on the dry background. These prints become fish when paper triangles are added for fins and when eyes and mouths are painted on.

Suggested Readings

Lionni, Leo. *Fish Is Fish.* New York: Pantheon, 1970.

Shaw, Evelyn. *Fish Out of School.* New York: Harper & Row, Pub., 1970.

Simon, Seymour. *Discovering What Goldfish Do.* New York: McGraw-Hill, 1970.

Wildsmith, Brian. *Fishes.* New York: Franklin Watts, Inc., 1968.

SHELLS

Information for Adults

Shell animals, or mollusks, make an interesting subject to study because children enjoy the beauty and novelty of the empty shells. They especially enjoy watching a live mollusk, which can become an unusual sort of pet. Mollusks are usually found near water, but they may also be found in inland areas. Mollusks are actually divided into six categories or classes, but for the children this may be simplified by using two main categories: univalves, or one-shell mollusks, and bivalves, or two-shell mollusks.

The category of univalves includes land snails and water snails, such as cone shells, conchs, welks, periwinkles, cowries, and many more. Univalves, which are one-shell mollusks, are also called gastropods, which means "stomach foot," because as these animals move along on their one foot it looks as if they're crawling on their stomachs. These animals protect themselves by retreating into their shells.

Bivalves, which are two-shell mollusks, include clams, scallops, oysters, cockles, and mussels. The two shells of a bivalve are hinged and can open and close through the aid of strong muscles. Some bivalves, such as the clam, move by burrowing with their one foot. The scallop moves by a sort of jet propulsion as it squirts water out of its shell with great force to push it along in the water. Some bivalves, such as oysters, attach themselves to an object early in their development and never move from that spot.

Shells offer protection for mollusks and give the animal shape, as the shell is a sort of external skeleton. Bivalves protect themselves by tightly closing their shells to keep intruders out. Shells are made by secretions that come from a sac covering called a mantle on the animal's body. The mantle covers the inner organs of the animal and is much like a substitute for skin. The semiliquid secreted by the mantle is made up of calcium and other minerals taken from the animal's food. These secretions, which are excreted unconsciously by the animal, harden as they come in contact with the air. On some shells, growth lines or rings may be found. A clam shell is a good example of one showing growth rings. The shape of a shell depends upon the shape of the mantle edge from which it was secreted. Shells are different colors and textures because mantles have different materials or minerals in their secretions. These differences are caused, in part, by what the animal has eaten as well as by what family the mollusk belongs to. These family differences may be compared to human heredity traits, such as hair, skin, and eye color; skin texture; and the size and shape of body parts.

Shells are often named for objects that they resemble, a fact that children find particularly amusing, as they like to try guessing the names of shells. Some shells popular with the children are slipper shells, butterfly shells, angel wings, turkey wings, worm shells, cone shells, and bleeding tooth shells. Conchs are popular because the children are able to place one ear against the opening of the shell and hear air circulating, which sounds like the roar of the sea. Cowries are also popular because a musical sound may be created by clinking two cowries together.

In the past, shells have been used for decorative purposes, as jewelry, musical instruments, and money; for obtaining dye; and even for a type of silk. Today, shell animals are enjoyed as food as well as for their beauty; however, many people consider mollusks harmful because of their destruction of vegetation on land and of boats and docks in the water.

In order to obtain a catalogue for ordering seashells, write to Tropical Shell Institute, Department BH-8, Box 2149, Ft. Lauderdale, Florida, 33335. Most of the books suggested at the end of this unit offer helpful information for identifying various seashells.

Experience: Part I

PURPOSE: To examine the color, shape, texture, and general appearance of a variety of shells and to learn the names for some of them.

To understand that animals once lived in each of the shells.

MATERIALS: A collection of seashells; a small piece of modeling clay; a picture of a mollusk inside its shell (or a live mollusk purchased from a store that carries aquarium supplies).

Display the collection of shells and let the children examine them. Tell them that many shells are named for what they look like. Provide names for the shells and see if the children can guess which of the shells is a cone shell, butterfly, turkey wing, slipper, bleeding tooth, and so on.

Ask the children if they can tell you what shells are for. If a child knows that each shell once housed a live animal, ask if they can take the piece of clay and show where the animal lived. Show the picture of the mollusk in its shell or let the children observe the live mollusk. Let each child take some of the clay to place in the shells so that they will begin to understand where the animal lived.

Suggested Questions

1. Which shell do you think is a cone shell? Which shell is a turkey wing shell?
2. What are shells for? What purpose do shells serve?

Experience: Part II

PURPOSE: To understand that a shell acts as a skeleton on the outside of the animal's body.

To understand that the mollusk's "skeleton," or shell, serves the same purpose as a human skeleton: It gives form and shape and offers protection.

MATERIALS: An oyster, clam, scallop, or other bivalve in the shell (from a fish market or some supermarkets).

Show the children the animal that lives inside the shell and let them examine it. Ask whether it has a shape that stays the same or whether it changes its shape as they handle it. Place the animal in one half of the shell and examine it again in order to see whether it has a definite shape that remains the same. Ask the children whether they think the animal could be easily crushed when outside its shell. Close the animal up inside both shells and ask the children to squeeze the shell in order to see if the animal could be easily hurt while inside the shell.

Suggested Questions

1. Does this shell animal have a definite shape when we handle it or is it just a jellylike mass that changes shape?
2. What does the shell do for the animal's shape?
3. Why would the animal be easily hurt when outside its shell?
4. How does a shell keep the animal from being hurt?
5. What would we be like if we had shells? How would our lives be different?

Experience: Part III

PURPOSE: To discover that mollusks have either one or two shells.

To observe the variety of colors found in shells.

To find the growth rings on the outside of some shells and to find the chamber or chambers inside some shells.

To discuss how univalves and bivalves move and how they protect themselves.

MATERIALS: A collection of shells (some univalves and some bivalves); two small boxes; a chambered nautilus or other univalve that has been cut open.

Have the children examine the collection of shells in order to determine how many shells mollusks have. When they have decided that some mollusks have one shell and some have two shells, the adult may wish to introduce the words univalve and bivalve. Let the children use the two boxes to categorize the shells into these two groups.

Ask the children for their ideas of why shells come in so many different colors. The adult may wish to explain how a mantle is formed and what influences the color of shells. The fact that shells in the same family may be of basically the same appearance may be compared to hereditary traits in humans. Help the children find a shell with growth rings.

Let the children explore the inside of the univalve that has been sawed open in order to see where the animal can retreat inside the shell when being attacked. Discuss what an animal with two shells might do in order to protect itself.

The adult may wish to discuss with the children how various shell animals move. The children may place one finger under a univalve shell in order to get an idea of how the animal looks as it crawls along. See "Information for Adults" for specific ways in which bivalves move.

Suggested Questions

1. How many shells do mollusks have?
2. Why are shells different colors and shapes?
3. If you could design a new shell animal, what would it look like?

Follow-Up Activities

Let the children go on a seashell hunt. If it is not possible to go to the seashore, scatter shells in the sand box and let the children hunt for them. After the shells have been collected, have the children name as many as they are able to, and then categorize the shells into univalves and bivalves.

Shells may be used as objects for printing with tempera paint on tissue paper or in a "spatter paint" design to create beautiful wrapping paper. To make spatter paint designs, arrange the shells on tissue paper. Dip an old toothbrush into tempera paint and scrape the wet bristles over the teeth of a comb to spatter paint over the surface of the tissue paper. Remove the shells to observe the prints they have left on the paper. Shells may also decorate small boxes or picture frames to be used as gifts.

Suggested Readings

Goudey, Alice. *Houses From the Sea.* New York: Scribner's, 1959.

Podendorf, Illa. *The True Book of Pebbles and Shells.* Chicago: Children's Press, 1960.

Posell, Elsa Z. *The Beginning Knowledge Book of Seashells.* New York: Crowell-Collier Press, 1969.

DAY AND NIGHT
THE PASSING OF TIME

Information for Adults

The earth is in a constant state of movement. As it orbits around the sun, it spins or rotates on an axis (an imaginary line running through the center of the earth extending from the North Pole to the South Pole). It is this rotation that creates day and night and helps to mark the passage of time. One complete rotation of the earth takes about twenty-four hours, marking the passing of an entire day and night. Three hundred and sixty-five days and nights make one calendar year, the approximate time it takes for the earth to make a complete orbit around the sun.

Long ago, people were satisfied with marking time merely in terms of days and nights, but eventually, it was discovered that the position of the sun could be used to mark significant times of the day. Later, the shadows created by the sun were used to mark time, and the first clock, a sundial, was invented. The daylight hours were divided into twelve equal parts, twelve hours. With no sun to mark the time during darkness, an hourglass was eventually developed to divide the night into twelve equal parts or hours.

The concept of day and night is relatively simple to explain to children if the earth's orbit around the sun is not explained at the same time. For the children, the rotation of the earth may be explained in terms of a dark side and a light side of the earth. If an imaginary line is drawn from the North Pole to the South Pole dividing the earth into two equal parts, one half, facing the sun, will be light, whereas the other half, away from the sun, will be dark.

The darkness of the night sky is the result of the absence of light. In order for color to be visible, there must be light, and black is the result of the absence of all light. The blue sky during the day is created by the scattering of sunlight by gaseous molecules in the air. Sunlight is composed of light of different wavelengths, and the blues created by shorter wavelengths are more easily scattered. The blue light goes in all directions and is reflected by water and dust particles in the air. Near the sun, the light is not as easily scattered, and all the colors of the spectrum are present. These colors combine to create a white appearance. When the sun is viewed at an angle less than vertical and when its light must pass through more of the earth's atmosphere, such as at sunrise or sunset, the longer red wavelengths, which are less easily scattered, are able to penetrate the atmosphere. The red light is scattered and reflected by the many particles in the air near the earth's surface. Often, the result is the colorful sunrises and sunsets that are visible.

Experience: Part I

PURPOSE: To discover what causes day and night.

MATERIALS: A high intensity lamp or projector; a globe or basketball; masking tape.

Discuss with the children that the only natural source of light is the sun. Tell them to think of the light source as the sun and the globe or basketball as the earth. In a darkened room, hold the ball or globe in the path of light from the lamp or projector as the children observe from the side. Ask them to find a line running through the center of the ball or globe that divides it into a light and a dark side. Discuss which side indicates day and which indicates night.

Place a piece of tape on the ball or globe to indicate the position of the children. Illustrate the rotation of the earth by slowly turning the ball or globe, so that the children are able to observe that they move from daylight to darkness, back to daylight, and so on. Explain dawn as a time when the children have been in darkness and are moving into the light. Dusk is a time when the children have been in light and are moving into darkness.

Suggested Questions

1. When we hold the ball or globe (the earth) in the path of light, are you able to see a dark side and a light side? What may we compare the darkness to? What may we compare the light to?
2. From where does the light that causes daylight come?

3. What causes us to move from light to darkness and back to light?
4. What would happen if the earth did not turn? Would we have day and night?

Experience: Part II

PURPOSE: To illustrate how the sun is used to tell the time of day.

MATERIALS: A half-circle cut from poster board and a marking pen; or chalk and a sidewalk; a stick that may be placed in the ground; a three-minute timer or an hourglass (optional).

On a sunny day, in an open area, place a half-circle of poster board on the ground or mark with chalk a half-circle on the edge of a sidewalk. Place a stick vertically in the ground next to the half-circle in the center of the straight edge. Beginning as early as possible in the morning, every hour on the hour (as determined by a clock), mark a straight line from the stick to the outer edge of the half-circle to indicate the shadow created by the stick. At the end of the day, add marks to indicate a shadow for the period of time, perhaps around sunrise, when you were unable to be present to mark the shadow created by the stick. The children will see that the marks indicating the hours are equally distributed around the half-circle to form half the face of a twenty-four-hour clock.

You may wish to complete the other half of the circle and indicate the other twelve hours of the day, so that the sundial resembles the face of a clock. Compare the sundial to a clock, discussing the fact that a clock indicates twelve equally divided hours instead of twenty-four.

If an hourglass or three-minute timer is available, you may wish to use them to discuss how people were able to tell time at night before clocks were invented.

Suggested Questions

1. What do we use to indicate the hours on the sundial? What creates the lines that indicate the hours?
2. When is the line created by the shadow of the stick the longest? The shortest?
3. How might you guess the time of day by looking at your own shadow?
4. Before there were watches or clocks, how were people able to tell time at night? Could they use a sundial?

Experience: Part III

PURPOSE: To demonstrate, in part, what causes the variations in the color of the sky.

MATERIALS: One flashlight; one clear drinking glass filled with water; one half teaspoon of milk.

Using "Experience: Part I" as a background, discuss with the children what causes night and why the sky is black at night. This may be demonstrated by moving into a closet or a room with no windows and turning out the lights. Ask the children what color they see in the room when there is no light.

In order to illustrate the reason for the blue of the sky during the day and the red, pink, or orange color at sunrise or sunset, shine the flashlight through the side of the glass of water and ask the children whether or not they see any colors in the water. Add one-half teaspoon of milk to the glass of water and stir. Shine the flashlight through the side of the glass of milky water in order to see that where the light penetrates directly, the liquid appears blue and that where it penetrates at an angle (at the surface), colors of pink, red, or orange appear.

Explain that the air is not really perfectly clear and that there are particles in it that change the sunlight, just as the milk in the water changed the light from the flashlight. When the sun shines directly through these particles during the day, the sky appears blue. At sunrise or sunset, when the sun shines through the particles at an angle, the sky appears red, pink, or orange.

Suggested Questions

1. What color do you see when there is no light? Why is the night sky black?
2. Did we add something to the water to make colors in the light appear? What colors did you see? Can you see these colors if there is no light?
3. What might cause colors to appear in the sky? From where does the light in the sky come?

Follow-Up Activities

For children who may have difficulty understanding the concept of night and day, it may be clarified by letting them stand in the position of the earth, in front of the bright light representing the sun. As they turn to face the light, they represent the side of the earth in daylight, and as they turn away from the light, they represent the side of the earth during nighttime.

The children may make individual sundials using a paper plate as the face of the sundial and a piece of clay to position a drinking straw (standing vertically) in the center of the plate. The hours may be marked on the plate using the shadow created by the straw.

Suggested Readings

Branley, Franklin. *What Makes Day and Night.* New York: Thomas Y. Crowell, 1961.

Brown, Margaret Wise. *A Child's Good Night Book.* New York: Young Scott Books, 1943.

Camden, Betty, and Green, Adolph. *Good Morning, Good Night.* New York: Holt, Rinehart & Winston, 1967.

Duvosin, Roger. *Day and Night.* New York: Knopf, 1960.

——. *Earth and Sky.* New York: Harper & Row, Pub., 1960.

Goudey, Alice. *The Day We Saw the Sun Come Up.* New York: Scribner's, 1961.

Ressner, Philip. *At Night.* New York: Dutton, 1967.

Zolotow, Charlotte. *Wake Up and Good Night.* New York: Harper & Row, Pub., 1971.

STARS

Information for Adults

Stars are glowing balls of gases generating heat like giant nuclear furnaces. The nearest star to the earth is the sun, which is ninety-three million miles away. Stars farther away differ in appearance from the sun because of their distance from the earth and because of the amount of heat they radiate. The sun appears to be larger than other stars because of its nearness, but it is actually a middle-sized star. Because the sun is so near the earth, the heat that it produces may be felt. This heat and the energy created by it is necessary in order to support life on the earth.

Stars glow brightly at all times. They cannot be seen during the day because the light from the sun floods the sky, making it so bright that the faint light of faraway stars cannot be seen. At times, stars cannot be seen at night because they are hidden by layers of clouds, and fewer stars may be seen even on a clear night if the moon is full. The best location to observe stars is away from the bright lights of a city.

Light from the stars travels through layers of clouds and moving air that surround the earth. The air and clouds bend and break the light, making the stars appear to blink off and on or twinkle. A "falling star" is not actually a star but a piece of rock or metal from outer space. This material burns brightly as it falls through space, creating the appearance of a star that is falling.

Astronomers learn about stars through the aid of a telescope, which helps them to observe stars that are too dim to see with the eyes alone, and through the aid of a spectroscope, which helps to determine how hot stars are, how fast they travel, and even what chemicals they are made of. Stars glow in different colors according to the amount of heat they radiate. Star colors are red, orange, yellow, white, and blue. Red is the color of the coolest star, and blue is the color of the hottest star. The sun, a yellow star, falls into the middle of the spectrum.

All the stars that may be seen from the earth and many that cannot, make up a group of stars called a galaxy. There are more than one-hundred billion stars in the earth's galaxy, and there are tens of billions of other galaxies in the universe. Long ago, people gave stars or groups of stars (constellations) names. The stars or constellations were named to commemorate great people or events or for animals or objects that the constellations resemble. Some stars in the earth's galaxy form a foggy band that has been named the Milky Way.

As the earth moves around the sun during its yearly orbit, the sun appears to pass through different groups of stars; therefore, different stars and constellations may be seen throughout the year. Stars appear to move across the sky or rise and set because of the earth's daily rotation. During this rotation, the North Star (Polaris) seems to remain in a nearly stationary position in the sky. The other stars appear to be moving or circling around Polaris.

People have long relied on stars to mark the seasons and the years and to guide them on long journeys. Stars, especially the North Star, are used to mark their course as they travel. Today, ships and airplanes are equipped with elaborate instruments for navigation; however, navigators of ships and military aircraft are still required to have some knowledge of the stars.

Experience: Part I

PURPOSE: To discover that stars do not "go off" or disappear during the day and to learn why they cannot be seen.

MATERIALS: A small flashlight; a source of very bright light, such as a film or filmstrip projector.

In a darkened room, turn on the flashlight and ask the children to think of it as a shining star. The children will be able to observe that when the light is shining in darkness, it may be easily seen. Now turn on the projector light and ask the children to think of it as the bright sunlight. With the flashlight still on, point it directly into the light of the projector (near the projector light). Ask the children to tell you whether the flashlight is still on. Turn off the projector light in order to show them that the flashlight is still on and could not be seen in the bright light of the projector. Discuss this illustration in terms of the stars and the sun.

Suggested Questions

1. Can you see the flashlight clearly in the darkness?
2. Can you see the stars clearly when it is dark (at night)?
3. What happened to the light of the flashlight when it was in the beam of the projector light? Was it still on, even though it couldn't be seen?
4. What is in the sky that produces such a bright light that the stars cannot be seen during the day?

Experience: Part II

PURPOSE: To discover what makes stars appear to twinkle.

MATERIALS: A piece of black construction paper; one straight pin for each child.

Let the children randomly punch holes in the black paper with their straight pins. When this paper is held in front of a bright light, it will now resemble the night sky filled with stars. Ask the children whether the "stars" on the paper are now twinkling. Have them move one hand between the back of the paper and the light in order to observe that the "stars" now appear to blink off and on or twinkle. Have children discuss what there is moving between the earth and the stars in the sky that could cause the stars to twinkle.

Suggested Questions

1. As the children move their hands between the black paper and the light, ask, "Are the 'stars' on your paper twinkling?"
2. What is causing the "stars" on your paper to twinkle? Do they twinkle when you hold your hand still?
3. What might be moving between the earth and the stars in the sky that could cause these stars to twinkle?

Experience: Part III

PURPOSE: To help the children learn what a constellation is and how to identify some constellations in the night sky.

MATERIALS: Black construction paper; a straight pin; a bright light or an overhead projector; the illustrations of constellations found in Figures 4–3a and 4–3b.

Place the constellation patterns in Figures 4–3a and 4–3b, one at a time, on top of a piece of black construction paper. Use the straight pin to punch a hole representing each star in the constellation (brighter stars should be indicated by larger holes).

The children may now hold the punched black paper in front of a bright light or place the paper on an overhead projector in order to project the constellation pattern on a screen

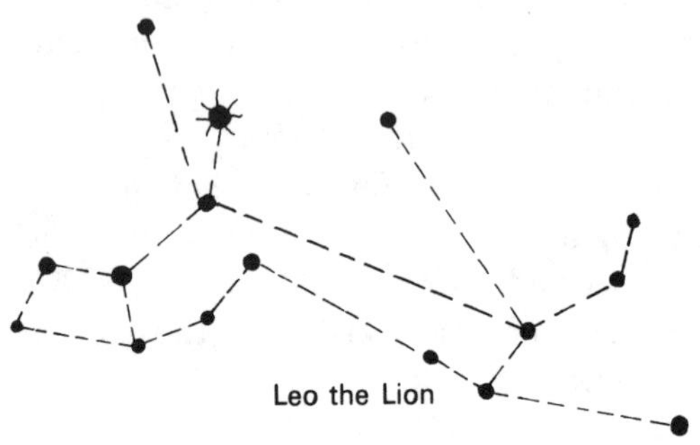

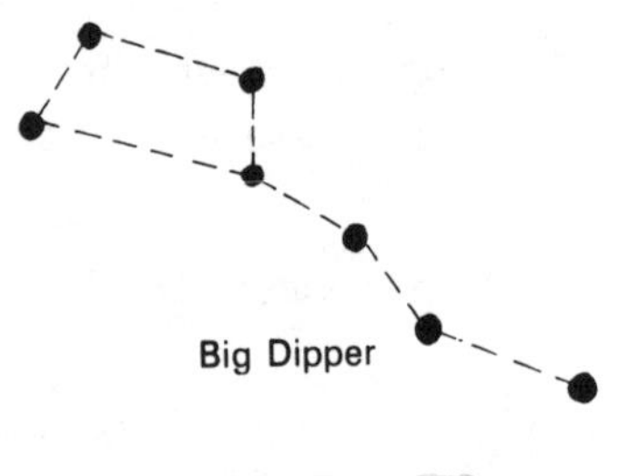

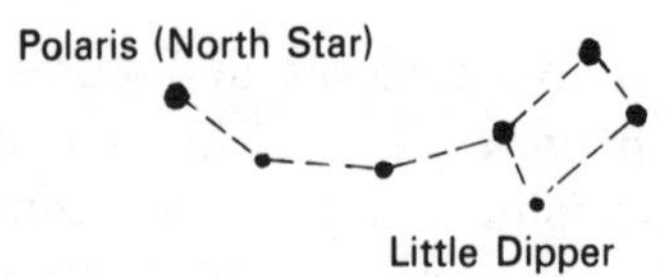

Figure 4-3a
Constellation Pattern: Spring Sky

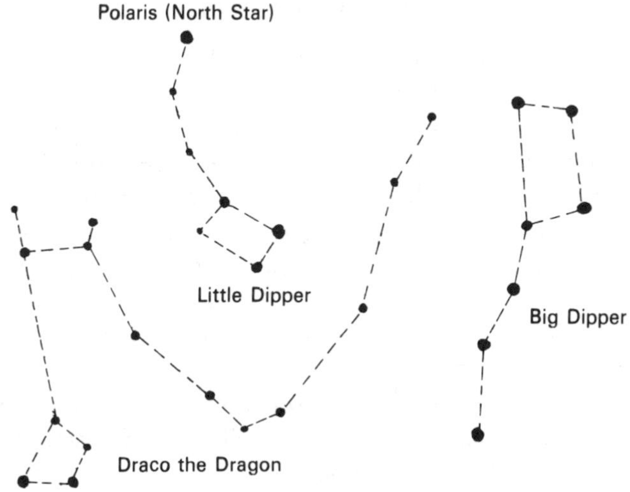

Figure 4-3b
Constellation Pattern: Summer Sky

or wall. Either method resembles the constellations in the night sky. Tell the children that these groups of stars are called constellations. Help them find Polaris (the North Star) and use it to locate the constellations. Let the children trace the pattern of each constellation with a finger and see if they are able to guess the name of the constellation from the design that it creates. Work with one constellation at a time until the children are able to locate each constellation in the night sky.

Suggested Questions

1. What does your paper look like when you hold it in front of the bright light? When we put the punched paper on the overhead projector, what does the screen look like?
2. What are these groups of stars called? Do the groups of stars look like an object, an animal, or a person? For what do you think constellations are named?

Follow-Up Activities

Let the children make individual planetariums. Each child will need one oatmeal box; one sharp object for punching holes; and one constellation pattern, with Polaris centered in the pattern. Tape the constellation pattern on the lid of the oatmeal box and let the children punch a hole to represent each star in the constellation. In the bottom of the box, cut a "peep hole" about one-quarter inch in diameter. As the child looks through the peep hole and slowly turns the lid at the other end of the box, the "constellations" will revolve around Polaris as they appear to do in the sky.

Suggest that as the children observe the stars, they regroup them in order to design their own constellations. Suggest that they give their new constellations names.

For a math experience, use constellations to create dot-to-dot drawings. Represent each star of a constellation with a dot, as in Figures 4–3a and 4–3b. Number the dots consecutively, beginning with 1, and let the children complete the outline that the constellation forms by drawing lines between each dot, beginning with number 1, moving to dot number 2, then to dot number 3, and so on. See if the children are able to guess the names of the constellations from the figures the stars create when connected with lines.

Listen to "Riding on a Star," from *Come and See the Peppermint Tree*, Dean Records, 2735 Macomb Street, N.W., Washington, D.C., 20088.

Suggested Readings

Berger, Melvin. *Stars.* New York: Coward, McCann & Geoghegan, 1971.

Branley, Franklin M. *The Sun: Our Nearest Star.* New York: Thomas Y. Crowell, 1961.

Kinney, Jean, and Kinney, Cle. *What Does the Sun Do?* New York: Young Scott Books, 1967.

Nussbaum, Hedda, editor. *Charlie Brown's Second Super Book of Questions and Answers.* New York: Random House, 1977.

Rockwell, Anne. *The Dancing Stars.* New York: Thomas Y. Crowell, 1972.

Schneider, Herman, and Schneider, Nina. *You Among the Stars.* Reading, Mass.: Addison-Wesley, 1961.

THE MOON

Information for Adults

The moon, like the sun and the earth, is a sphere; its shape remains constant. As the moon orbits the earth, its shape does appear to change as its relationship to the sun and earth changes. During its orbit, the moon appears to grow larger (a waxing moon) until it is full, and then it appears to get smaller again (a waning moon) until the next new moon. One complete cycle takes about twenty-nine to thirty days and is termed a lunar month.

The shapes of the moon are called phases, and they include the following: new moon (often called no moon because no moon is visible), crescent, quarter moon (sometimes called a half moon, because the moon appears to be a half-circle), gibbous (a moon that does not appear as a full circle but bulges or curves on all sides), and full moon (when the moon appears as a full circle). See Figure 4-4. Times given for the rising of the moon are approximate.

Because of the earth's rotation on its axis, the moon appears to rise and set much as the sun does. However, because of the moon's monthly orbit around the earth, its position in relation to the earth changes daily. Therefore, the moon rises and sets at a later time each day (approximately fifty minutes later) as it moves from one new moon to the next. This makes it possible to see the moon during the day at certain times in the lunar month (just before or after the new moon).

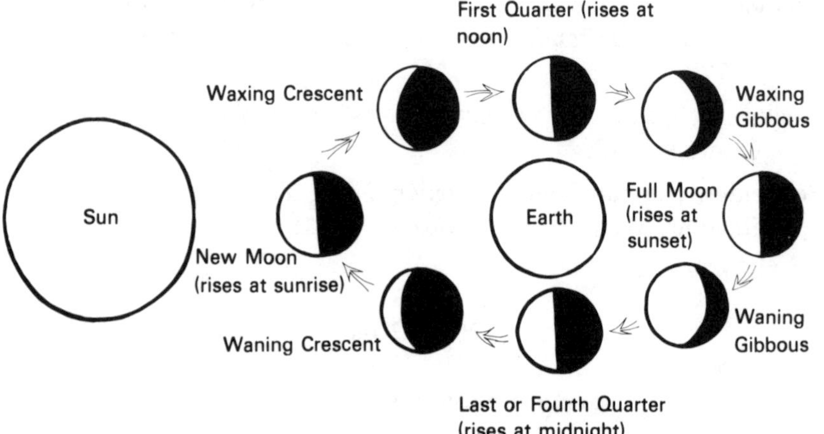

Figure 4-4

Since the moon produces no heat or light of its own, the light called moonlight is actually sunlight being reflected off the moon. The surface of the moon is like a barren desert with dust, rocks, mountains, and craters. Viewed from the earth, shadows produced by the mountains and craters give the illusion of a face—the Man in the Moon.

Experience: Part I

PURPOSE: To become aware of the various phases of the moon.

To learn to identify and to name some of the phases of the moon.

MATERIALS: Cardboard shapes representing the phases of the moon; a large piece of poster board; one magic marker.

Hold up each shape representing a phase of the moon and ask the children whether they have ever seen the moon when it looked like this shape. Give them the name for each shape or phase of the moon.

With poster board, make a large calendar to show the days of the lunar month, beginning with the new moon and ending with the next new moon. The lunar month may cover parts of two calendar months, so mark each square of the calendar with the month and the day of the month, and mark the phases of the moon with a picture of the appropriate shape for that date. Many calendars and newspapers provide information on the dates of the phases of the moon. Tell the children the hours when they may best see the moon each day or night. After they have observed the moon in the sky, let them choose a cardboard shape that matches the shape of the moon that they saw.

Suggested Questions

1. Have you ever seen the moon when it looked like this shape?
2. Can you show me the shape that is the same as the shape of the moon when you saw it in the sky?

Experience: Part II

PURPOSE: To discover why the moon appears to change shape.

MATERIALS: A very dark room; a projector, a lamp, or strong flashlight; a large solid color ball; cardboard shapes representing the phases of the moon.

This experience must be presented individually in order for each child to see what causes the changing phases of the moon. The light source should be placed at a height that is just above the child's head. Mark an "x" on the floor three to four feet in front of the light and have the child stand on it. Have the child think of himself or herself as being on the earth, of the ball as the moon, and of the light as the sun. You should hold the ball and walk around the child in order to show the moon's orbit around the earth. Repeat, but stop at each position that illustrates a phase of the moon (from the child's position of observation). Let the child trace with a finger the shape that the light creates on the ball, and then find a cardboard shape that matches the shape or phase that the light has created on the ball (the moon). Ask if the child knows the name for this phase of the moon.

Suggested Questions

1. Does the sun shine on the moon in different ways as the moon travels around the earth?
2. How does the moon (ball) look when it is between you and the sun?
3. Does the ball have any light of its own when the light is not shining on it?
4. From where do you think the moon gets light?

Figure 4-5
Moon Mobile

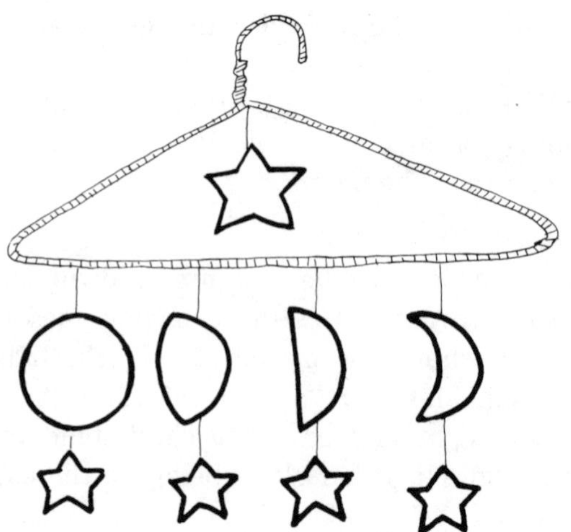

Follow-Up Activities

For an art activity, let the children make a "moon mobile," using shapes that represent the phases of the moon from the night sky (see Figure 4-5).

Listen to "The Moon in the Yard" from *Come and See the Peppermint Tree,* available from Dean Records, 2735 Macomb Street, N.W., Washington, D.C., 20088.

Suggested Readings

Branley, Franklin. *The Moon Seems to Change.* New York: Thomas Y. Crowell, 1960.

Brown, Margaret Wise. *Goodnight Moon.* New York: Harper & Row, Pub., 1947.

BIBLIOGRAPHY

Adler, Irving. *Electricity in Your Life.* New York: Harper & Row, Pub., 1965.

Allen A., Ph.D. *The Book of Bird Life.* New York: D. Van Nostrand Company, 1961.

Baker, Jeffrey J. *The Vital Process: Photosynthesis.* Garden City, N.Y.: Doubleday, 1969.

Bendick, Jeanne. *Adaptation.* New York: Franklin Watts, Inc., 1971.

Blough, Glen. *Bird Watchers and Bird Feeders.* New York: McGraw-Hill, 1963.

——. *After the Sun Goes Down.* New York: McGraw-Hill, 1956.

Branley, Franklin. *The Moon Seems to Change.* New York: Thomas Y. Crowell, 1960.

——. *Color: From Rainbows to Lasers.* New York: Thomas Y. Crowell, 1978.

Brouillette, Jeanne. *Insects*. Chicago: Follett Publishing Co., 1963. 1963.

——. *Moths*. Chicago: Follett Publishing Co., 1966.

——. *Butterflies*. Chicago: Follett Publishing Co., 1961.

Buck, Margaret Waring. *Where They Go in Winter*. New York: Abington Press, 1968.

Burns, Aubrey. *Bird Watching*. New York: Franklin Watts, Inc., 1968.

Byers, Horace Robert. *General Meteorology*. New York: McGraw-Hill, 1959.

Callahan, Philip S. *Bird Behavior*. New York: Four Winds Press, 1975.

Cavanna, Betty. *The First Book of Wildflowers*. New York: Franklin Watts, Inc., 1961.

Chandler, T. J. *The Air Around Us*. Garden City, N.Y.: The Natural History Press, 1969.

Cloudsley-Thompson, J. L. *Spiders and Scorpions*. New York: McGraw-Hill, 1974.

Darling, Lois, and Darling, Louis. *Worms*. New York: Morrow, 1967.

Dittmer, Howard. *Modern Plant Biology*. New York: Van Nostrand Reinhold, 1972.

Echard, Margaret. *Hoofs, Paws, and Hands*. San Carlos, Calif.: Golden Gate Junior Books, 1968.

Editors of Creative Educational Society. *Geology of the Earth*. Mankato, Minn.: Creative Educational Society, Inc., 1971.

Elgin, Kathleen. *The Human Body and the Skin*. New York: Franklin Watts, Inc., 1970.

——. *The Human Body: The Muscles*. New York: Franklin Watts, Inc., 1973.

Fishlock, David. *Taking the Temperature*. New York: Coward, McCann & Geoghegan, 1968.

Fox, Charles Philip. *When Winter Comes.* Chicago: Reily & Lee Co., 1962.

Gates, Richard. *The True Book of Conservation.* Chicago: Children's Press, 1959.

Gromer, Richard. *Soil.* Chicago: Follett Publishing Co., 1967.

Harvey, Fran. *Why Does It Rain?* Irvington-on-Hudson, N.Y.: Harvey House, Inc., 1969.

Hirsh, S. Carl. *The Living Community.* New York: Viking, 1966.

Hutchins, Ross E. *The Amazing Seeds.* New York: Dodd, Mead, 1960.

——. *Hop, Skim, and Fly.* New York: Parent's Magazine Press, 1970.

——. *The Bug Clan.* New York: Dodd, Mead, 1973.

Irving, Robert. *Volcanoes and Earthquakes.* New York: Knopf, 1962.

Jensen, Amy Elizabeth. *Mushrooms, Ferns and Mosses.* New York: Grosset & Dunlap, 1965.

Knight, David C. *Let's Find Out About Weather.* New York: Franklin Watts, Inc., 1967.

Lieberg, Owens. *Wonders of Heat and Light.* New York: Dodd, Mead, 1966.

List, Ilka. *Questions and Answers About Seashore Life.* New York: Four Winds Press, 1970.

Lubell, Winifred. *In a Running Brook.* Chicago: Rand McNally, 1968.

May, Julian. *Why Birds Migrate.* New York: Holiday House, 1970.

McClung, Robert M. *Moths and Butterflies and How They Live.* New York: Morrow, 1966.

McFall, Christie. *Wonderful Stones.* New York: Dodd, Mead, 1962.

Milgram, Harry. *Understanding Weather.* New York: Crowell-Collier Press, 1970.

Owen, S. *Wonders of Magnets and Magnetism.* New York: Dodd, Mead, 1967.

Page, Lou Williams. *Rocks and Minerals.* Chicago: Follett Publishing Co., 1962.

Pringle, Laurence. *Ecology: Science of Survival.* New York: Macmillan, 1971.

Reid, Keith. *Nature's Network.* Garden City, N.Y.: The Natural History Press, 1970.

Rosenfeld, Sam. *Ask Me a Question About the Weather.* Irvington-on-Hudson, N.Y.: Harvey House, Inc., 1966.

Russell, Helen Ross. *Winter: A Field Trip Guide.* Boston: Little, Brown, 1972.

Sarasy, Phyllis. *Winter Sleepers.* Englewood Cliffs, N.J.: Prentice-Hall, 1969.

Schima, Marilyn. *The Magic of Life.* Englewood Cliffs, N.J.: Prentice-Hall, 1970.

Schlichting, Harold E., Jr. *Ecology: The Study of Environment.* Austin, Tex.: Steck, Vaughn, Co., 1971.

Schneider, Herman, and Schneider, Nina. *You Among the Stars.* Reading, Mass.: Addison-Wesley, 1961.

Selsam, Millicent E., and Hunt, Joyce. *A First Look at Birds.* New York: Walker and Co., 1973.

Stambler, Irwin. *The Worlds of Sounds.* New York: W. W. Norton & Co., Inc., 1967.

Stone, A. Harris, and Siegel, Bertram M. *Turned On: A Look at Electricity.* Englewood Cliffs, N.J.: Prentice-Hall, 1970.

Swain, Ralph B. *The Insect Guide.* Garden City, N.Y.: Doubleday, 1948.

Tellander, Marian. *Space.* Chicago: Follett Publishing Co., 1960.

Wasson, Isabel B. *Birds.* Chicago: Follett Publishing Co., 1963.

Weart, Edith Lucie. *The Story of Your Bones.* New York: Coward, McCann & Geoghegan, 1966.

——. *The Story of Your Respiratory System.* New York: Coward, McCann & Geoghegan, 1964.

Woods, Loren. *Fishes.* Chicago: Follett Publishing Co., 1969.

Wyler, Rose, and Ames, Gerald. *Secrets in Stones.* New York: Four Winds Press, 1970.

Zim, Herbert S. *Goldfish.* New York: Morrow, 1947.

APPENDIX I

Dinosaurs

BRONTOSAURUS (Bron'-toe-saur'-us). The name means "thunder lizard." Twice the length of an average-sized classroom, sixty to seventy feet long, and weighing eighty thousand pounds (more than the weight of ten elephants), the Brontosaurus had a proportionately small head and brain. This dinosaur was a docile, plant-eater that spent most of its life wading in lakes or swamps. The water supplied a source of food and helped to support the weight of its enormous body.

TYRANNOSAURUS REX (Tie-ran'-o-saur'-us rex). Called the "king of tyrants," the Tyrannosaurus was the most ferocious and terrifying animal ever to walk the earth. About twenty feet high and fifty feet long, it walked on its hind legs. The two front legs were so short they could not reach its mouth, but the back feet had sharp, tearing claws like those of an eagle. The mouth could open four feet wide and had rows of razor-sharp teeth six inches long. Being a meat-eater, it preyed on plant-eaters.

STEGOSAURUS (Steg'-o-saur'-us). An armored dinosaur measuring fifteen to twenty feet long and weighing seven or eight tons (larger than a large elephant), the Stegosaurus had a hard shell or plating of armor for protection from the meat-eaters. On its back were triangular-shaped bony plates as much as two feet tall. On the end of its tail were spikes three feet long. As the Stegosaurus would swing its tail, the spikes could rip open its enemies, the meat-eating dinosaurs. A plant-eater, the Stegosaurus had a short neck and hind legs much longer than its short front legs; it walked on all fours, with its back end in the air.

TRACHODON (Track'-o-don). The name means "rough tooth." There were no teeth in the front of its mouth, but in the back jaws were rows of teeth piled together. These teeth had a very bumpy or rough appearance, like that of a cobblestone street; thus the name, "rough tooth." About thirty feet long and sixteen to eighteen feet high, the Trachodon could walk on all fours or just on its hind legs. The body was shaped somewhat like that of a kangaroo, but the face was shaped much like that of a duck, with a broad, flat bill suitable for eating only plants. The Trachodon protected itself by escaping into the water and using its webbed front feet and heavy flattened tail as swimming aids.

TRICERATOPS (Tri-cer'-a-tops). An armored dinosaur with three horns, two extending from the head and one from the nose, the Triceratops vaguely resembled a rhinoceros with horns and a collar. The collar was a shield around its neck providing protection in the spot where meat-eaters were most likely to attack. Measuring about thirty feet long and weighing about twelve tons, the Triceratops was a plant-eater and an excellent fighter.

APPENDIX II

A Story About Pollution*

Once there was a beautiful town named Cleantown. Cleantown was a wonderful place to live. It had many beautiful trees and flowers. There were parks with lots of grass and clear, clean lakes

*Adapted by permission of Scholastic Book Services, a division of Scholastic Magazines, Inc. from *Scholastics Earth Corps Environmental Awareness Series: Look Around You* by Wendy Cheyette Lewison and Terry Ann Simon, copyright © 1971 by Scholastic Magazines, Inc.

and streams all around. The streets were clean and lined with pretty houses that had yards for the children to play in. The people in Cleantown were very happy and smiled a lot. They said "hello" to each other as they walked along in the clean, fresh air.

Soon, other people began to hear about Cleantown, and they wanted to live there, too. Many, many people began moving to Cleantown. The people chopped down all the trees and dug up the grass and flowers in order to make room to build more streets and many houses and factories all crowded together. They drove cars and trucks with dirty smoke coming out of the back, and they honked their noisy horns. They began throwing litter into the streets and dumping waste into the water.

Soon, the air was dirty with smoke and smog, and the streets and streams were dirty with litter and waste. The people stopped smiling, and they didn't say "hello" to each other as they passed on the street. They did not like what had happened to their town, and they wanted to make it better again. Can you tell them what they should do to turn their town back into "Cleantown"? They need your help.

APPENDIX III

Insects

ANTS. Ants are termed social insects and live in a group called a colony. The colony functions much like a family, with individuals cooperatively caring for the young, division of labor among workers, and an overlapping of two or more generations. Within the home or nest, there are many workers performing their own specific jobs. Some secure food; some care for the young; some construct the nest and dig the tunnels; and others keep it clean. The female ants are the workers; the males are present merely for mating and die after the queen has been fertilized. Each nest has one or more queens to lay the eggs. Ants are extremely strong for their size and are able to lift as much as four hundred and fifty times their weight.

This would be comparable to a child lifting three large cars at the same time. Ants communicate with each other using their antennae and chemical odors. In the winter, they dig deeply into the soil to find warmth. They protect themselves by escaping and by biting, and they are essentially helpful insects because they clean up the garbage humans leave behind.

BEES. Not all bees are social insects. Those that are live in groups or colonies and may reside in hives, each performing specialized jobs similar to those performed by ants. The workers are females, and each colony has a queen for laying eggs. The males, called drones, are present for mating purposes. Honeybees communicate through dances that indicate sources of nectar and its location. Bees are considered helpful for their contribution of honey, which is made from a combination of nectar, sipped through a strawlike tongue, and ingredients from the worker bee. Their most valuable contribution is the pollination of flowering plants. As the bees gather nectar, pollen from the flower is collected in pollen baskets on their hind legs and is carried from flower to flower. The bees mix some of the pollen with nectar in order to make "bee bread" for the young bees. Bees "air condition" their hives in the summer by stationing workers at the entrance to circulate the air by fanning their wings. The hive may be sealed off with wax in the winter, and the bees huddle together inside for warmth. Bees protect themselves by flying away to escape their enemies and by using a stinger on the rear of the body. If the stinger is lost, the bee will die.

BEETLES. Beetles are the most prevalent of all insects. Their outstanding characteristics include large jaws used for chewing and a pair of heavy wings that act as a shield for the flying wings underneath. The beetle order includes the misnamed ladybug and June bug. The ladybug beetles, popular with children, are not all females. Another popular, but misnamed, beetle is the firefly or lightning bug. This insect demonstrates a fluorescent appearance in order to attract mates. This ability to "light up" comes from a chemical, called luciferin, in the body. When this chemical comes in contact with oxygen (the amount being controlled somewhat by the insect), a heatless light is created. Beetles eat meat, decaying plant matter, and growing plants (this group would be considered pests). Beetles spend the winter wherever warmth can be found: under leaves, rocks, logs, and tree bark. Their main sources of protection are their hard shells and their ability to fly.

BUGS. Bug is a commonly misused term. All bugs are insects, but not all insects are bugs. The bug order is characterized by piercing and sucking mouth parts, an incomplete metamorphosis, and, generally, two sets of wings (usually, the front set of wings has a thickened or leathery portion at the base). Some common bugs include cicadas, water striders, frog-hoppers, spittle bugs, tree-hoppers, leaf-hoppers, aphids or plant lice, stink bugs, cabbage bugs, squash bugs, and kissing bugs. Some misnamed bugs include June bugs, ladybugs, lightning bugs, and potato bugs. Bugs are generally not considered desirable insects because of their destruction of plants. For protection, some bugs use camouflage, some hide, and some escape by flying away.

BUTTERFLIES AND MOTHS. These insects are interesting to the children because of their beauty and their life cycle, called a complete metamorphosis. This life cycle includes four stages: the egg, the larva or caterpillar, the pupa or cocoon (of a moth) or chrysalis (of a butterfly), and the adult. Butterflies and moths may generally be distinguished from each other in several ways. The moths may have featherlike antennae that do not end in knoblike structures; the butterfly has clublike antennae that always have knobs on the end. The moth usually has a thicker body and fuzzy wings; the butterfly is usually more slender, with smoother wings. At rest, the butterfly holds it wings vertically over its head, whereas the moth holds it wings outstretched. The time which the butterfly or moth spends in the pupal stage and the time at which the adult emerges depend upon the variety. Generally, moths spend the winter in a cocoon and emerge in the spring, and the caterpillars of butterflies spin a silk chrysalis in the spring. The adult of both, except for the Monarch butterfly, usually dies in the winter. The main source of protection for these insects is flying. Some display various methods of camouflage.

DRAGONFLIES AND DAMSELFLIES. These relatives are generally considered helpful insects because of their consumption of flies and mosquitoes. They catch their prey by forming a basket with their front feet. Their home is near the water where eggs are laid. The babies are called nymphs and remain in this stage for three years. The adult lives just one season and dies in the winter. The main source of protection for the adult is escape by flying.

GRASSHOPPERS, CRICKETS, AND KATYDIDS. This is the musical family of insects and includes insects with ears. It is the male that creates the music: Grasshoppers make music by rubbing their legs against their wings, and crickets and katydids make music by rubbing their wings together. The music is a mating call and, generally, may be heard only when it is dark. These insects consume grain and the leaves of young plants and are thus considered pests by farmers. Their main source of protection is camouflage and the ability to hop and fly. The insect generally lays eggs on the ground in the late fall, and the adult dies during the winter.

HOUSEFLIES AND MOSQUITOES. These relatives are found almost everywhere and are generally considered a nuisance, as they are disease carriers. Flies feed and breed on rotting plants and animals. The babies or larvae are called maggots. Their only form of protection is flying. Mosquitoes have piercing and sucking mouths and feed on animal blood. The babies are called wrigglers and are found in stagnant water. Mosquitoes protect themselves by flying away to escape from their enemies.

PRAYING MANTISES AND WALKING STICKS. These insects are related to the grasshopper but are considered helpful because of their vast consumption of insect pests. The praying mantis is considered the terror of the insect world. It sits in wait for its prey, catching it with its front legs. These insects may be tamed as pets and can learn to drink from a trough. The mantis and the walking stick use camouflage as their main source of protection. The mantis is able to fly, but the walking stick has no wings. The adult lays eggs in the late fall and dies. The eggs hatch in the spring.

WASPS. These insects are not generally classified as either helpful or harmful. Some have a painful sting, but they are also helpful in that some feed on caterpillars or are parasites on insect larvae and eggs. Most wasps are solitary, although a few are social. They live in a variety of houses. Some live in a type of paper house called a nest. It is composed of chewed bits of wood mixed with saliva. The nests may consist of cells resembling a bee's honeycomb. These cells house the unborn wasps. Mud daubers build their nests of mud on the sides of houses, under the eaves, and so on. Wasps protect themselves by stinging or flying. Workers are killed by the winter cold, but queens generally survive the winter.